I0131548

HARNESSING PRODUCTION NETWORKS

The **Institute of Southeast Asian Studies (ISEAS)** was established as an autonomous organization in 1968. It is a regional centre dedicated to the study of socio-political, security and economic trends and developments in Southeast Asia and its wider geostrategic and economic environment.

The Institute's research programmes are the Regional Economic Studies (RES, including ASEAN and APEC), Regional Strategic and Political Studies (RSPS), and Regional Social and Cultural Studies (RSCS).

ISEAS Publications, an established academic press, has issued more than 2,000 books and journals. It is the largest scholarly publisher of research about Southeast Asia from within the region. ISEAS Publications works with many other academic and trade publishers and distributors to disseminate important research and analyses from and about Southeast Asia to the rest of the world.

HARNESSING PRODUCTION NETWORKS

Impacts and Policy Implications from Thailand's Manufacturing Industries

AEKAPOL CHONGVILAIVAN

ISEAS
INSTITUTE OF SOUTHEAST ASIAN STUDIES
Singapore

First published in Singapore in 2011 by ISEAS Publishing
Institute of Southeast Asian Studies
30 Heng Mui Keng Terrace
Pasir Panjang
Singapore 119614

E-mail: publish@iseas.edu.sg
Website: http://bookshop.iseas.edu.sg

ISEAS Library Cataloguing-in-Publication Data

Aekapol Chongvilaivan.
Harnessing production networks : impacts and policy implications from Thailand's manufacturing industries.
1. Manufacturing industries—Thailand.
2. Industrial productivity—Thailand.
3. Labour productivity—Thailand.
4. Skilled labour—Thailand.
5. Contracting out—Thailand.
I. Title.
HD9736 T42A24 2011

ISBN 978-981-4311-26-7 (soft cover)
ISBN 978-981-4311-27-4 (E-book PDF)

Typeset by International Typesetters Pte Ltd
Printed in Singapore by Markano Print Media Pte Ltd

Contents

List of Tables and Figures

Acknowledgements

This book is a non-technical collection of my past research studies pertaining to the impacts of and policy implications on rising production networks which have emerged as a key production platform in the world of globalization. It has enormously benefitted from the intuitive insights of and discussions with Associate Professors Shandre M. Thangavelu and Jung Hur, both of whom inspired me to pursue in-depth research on this subject. I would also like to confer my special thanks to the Director of ISEAS, Ambassador K. Kesavapany; former Deputy Director of ISEAS, Dr Chin Kin Wah; Managing Editor of ISEAS Publications Unit, Mrs Triena Ong, and my colleagues at the Publications Unit, for their excellent support and suggestions.

Lastly, the completion of this book would not have been possible without the support and encouragement from my parents and my wife, Miyuki Chan. To them, I hereby devote this book.

1
Introduction

Production fragmentation is a phenomenon of great worldwide interest. Due to technical advances in both production and communication technology, fragmenting and contracting out production of intermediate materials and services have spilled over into the Southeast Asian arena. Even though fragmentation which has become part and parcel of manufacturing production has stimulated keen research interest in its economic consequences, there have been loopholes in the literature. First of all, most existing research on the subject is primarily concerned with the issues related to the industrialized countries. The economic impacts on developing Southeast Asia, however, has received only little attention. Second, despite countless reports on the magnitude and growth of production fragmentation, the distinctive nature of production fragmentation in developing economies like those in the Southeast Asian region makes thorough understandings of its impacts greatly subtle. Last, and perhaps most importantly, the absence of rigorous research is largely attributable to the lack of solid evidence that hallowed a continuous public debate without a strong information base.

This book attempts to bridge these gaps in the existing studies. In contrast with the existing literature, this book provides complete, yet non-technical, analyses of production fragmentation effects and thus targets a wide range of readers — including academics, researchers, policy-makers, students, entrepreneurs, and anyone who is interested in this subject. It investigates the economic impacts of production fragmentation, such as its effects on firms' performance, wage inequality and skills upgrading in Southeast Asia with a focus on Thailand's experience of emerging as a global hub of fragmentation and outsourcing. This book studies these phenomena from the perspectives of developing countries and thus elucidates new evidence in connection with production, industrial organization and labour economics theories, providing interesting insights for formulating industrialization and labour development policies.

Usual caveats apply. Although a number of issues such as measurement errors, causality problems, potential endogeneity biases, and modest goodness of fit statistics are by all means indispensable, they remain unanswered in my analyses, and therefore my findings should be contemplated as tentative. Tackling these issues requires extensive discussions and analyses and hence may be far beyond the scope of this book. I leave them for the possible avenues of future research.

1.1 A Primer on Production Fragmentation in Thailand's Manufacturing Industries

Throughout this book, the analytical framework employs establishment-level data of Thailand's manufacturing industries, namely the *2003 Manufacturing Industry Survey*, provided by the National Statistical Office (NSO) of Thailand. This dataset is a countrywide survey that contains detailed cost and production information, in which approximately 8,000 manufacturing establishments were engaged.

Notwithstanding the absence of comprehensive sources of data on outsourcing in the Southeast Asian region, anecdotal evidence suggests that relatively advanced countries in the region, including Thailand, are emerging as a hub of outsourcing activities, especially IT-related outsourcing, thanks to the globalization trend and increasingly competitive business environment. The proliferation of production fragmentation in the region creates new avenues for employment, productivity enhancement, and economic growth.

Since the late 1990s in the aftermath of the Asian financial crisis, the Thai economy has exhibited persistent surges in employment accompanied by expansions of manufacturing activities as portrayed in Figure 1.1. Tables 1.1 and 1.2 paint a more apparent picture of the magnitude of the manufacturing sector. The figures reveal that manufacturing value-added accounted for more than 33 per cent of Thailand's GDP during the period of 2002–06, and the manufacturing sector employed more than 5 million workers during the period of 2004–06 — nearly 67 per cent of which pertained to small and medium enterprises (SMEs). This implies that the manufacturing sector has been a key driving force of economic growth in Thailand in terms of both GDP contribution and employment.

FIGURE 1.1
Import, Employment and Manufacturing Indices (2000 = 100)

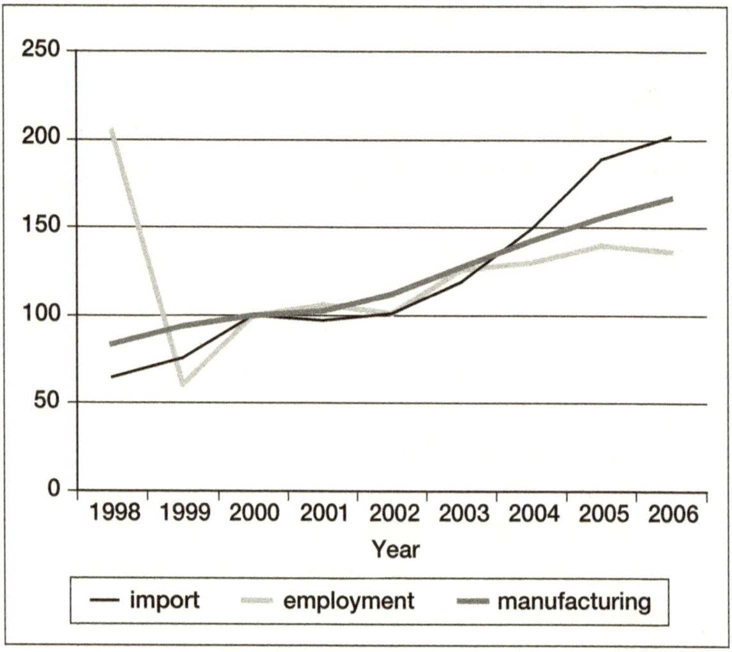

Source: Bank of Thailand.

Thailand has emerged as a natural hub of outsourcing activities due to its well developed infrastructures and proximity to Asia's biggest markets. Geographically, Thailand is located at the strategic location at the centre of Asia's burgeoning markets. Its transportation and telecommunication infrastructures such as international airports, internal waterways, railways, highways, deep sea ports, bandwidth and underwater cable result in low transportation costs, efficient delivery of goods and exceptional

TABLE 1.1

Manufacturing Value-Added (2002–06)

GDP (Million Baht)	*2002*	*2003*	*2004*	*2005*	*2006*
Whole Kingdom	5,450,643	5,917,368	6,489,847	7,087,660	7,816,474
Manufacturing	1,836,083	2,061,572	2,238,222	22,466,180	2,739,534
Manufacturing Share (%)	33.7	34.8	34.5	34.8	35.0

Source: *Report on 2006 Small and Medium Enterprises Situation and Outlook for* 2007, Office of Small and Medium Enterprises Promotion, Thailand.

TABLE 1.2

Manufacturing Employment (2004–06)

No. of Persons	*2004*	*2005*	*2006*
Manufacturing	5,143,277	5,193,482	5,228,190
SMEs	3,431,553	3,459,096	3,496,202
Share of SMEs (%)	66.7	66.6	66.9

Source: Office of Small and Medium Enterprises Promotion, Thailand.

access to supply and distributional networks. In addition, Thailand is also well endowed with a large pool of relatively low cost skilled workers. This enables outsourcing manufacturers to reap benefits from labour cost savings and specialization in core-competent activities.

Recent developments, nevertheless, indicate that the competitiveness of Thailand's manufacturing sector has deteriorated due to the rising domestic wages and prices on top of external factors such as a rise of huge emerging countries with a large pool of cheap labour like China and India. To secure their competitiveness in the global market, local manufacturers have increasingly fragmented and contracted out their production activities to outside providers, so-called production fragmentation or outsourcing, in order to reap benefits from more efficient operations. In the plastic industry, for instance the research and development (R&D) activities are prevalently contracted out due to the lack of expertise in pushing forward indigenous technology and in-house human capital. The textile industry likewise commonly outsources marketing and/or packaging activities to outsourcing providers to reduce production costs and improve specialization, thus staying competitive in the market.

As a result, production fragmentation that ties firms through vertical business linkages has attracted more attention from entrepreneurs, policy-makers and academics and by and large constituted a crux of the Thai industrial policy formation. Now that deepening trade and investment liberalization, coupled with rapid development and dissemination of information and communication technology (ICT), has made the global business environment in which firms operate greatly competitive, production fragmentation in Thailand's manufacturing industries is now not only limited to low skill-intensive, low value-added production activities like assembling and packaging, but also pertinent to sophisticated, high value-added activities such as R&D, information technology (IT) services, and product designs. As posited by Wattanapruttipaisan (2002), production fragmentation benefits Thai manufacturers in a number of ways. These include skill upgrading, standards and quality enhancement, production capacity development, managerial and organizational improvements, and outside financial source abundance, in addition to new know-how and technology firms obtain from linkages with outsourcing providers, especially those associated with multinational enterprises (MNEs) (Schmitz and Knorringa 2000).

1.2 Outsourcing Measurement

Despite the various accounts on this subject, production fragmentation for the time being is loosely coined. It may also be termed as "outsourcing" (Feenstra and Hanson 1996*a* and 1996*b*), "contracting out" (Abraham and Taylor 1996), "vertical disintegration" (Holmes 1999), "fragmentation" (Arndt and Kierzkowski 2001), "vertical specialisation" (Hummels et al. 2001), and a "make-or-buy decision" (Grossman and Helpman 2002). Throughout this book, I utilize these terminologies interchangeably.

The definition of outsourcing is less clear-cut. According to Olsen (2006), outsourcing broadly refers to "the relocation of jobs and/or processes to external providers regardless of the provider's location". In this sense, outsourcing may therefore be concerned with task relocations within or between countries or both.

In fact, two strands of literature investigating economic impacts of outsourcing have arisen. The first strand takes the view that the prevalence of outsourcing activities observed in the world is driven largely by the declines in transaction costs in connection with the intensified uses of information technology (Abraham and Taylor 1996; Girma and Görg 2004; and Díaz-Mora 2008, among others). Its main research inquiries weigh in on how a decision on contracting out business activities alters intra-firm productivity, performance and production efficiency. Therefore, it does not distinguish between domestic and international outsourcing. As argued by Girma and Görg (2004), since the subsequent productivity effects are of their interests, it should not matter whether outsourcing takes place internationally or domestically.

In contrast, the other strand is related specifically to the trade-related aspect of outsourcing since it deems international outsourcing as a key mechanism through which low skill-intensive production activities are moved from skilled labour-abundant countries, like the United States (U.S.), the European Union (EU) and Japan, to unskilled labour-abundant countries like developing Southeast Asia, China and India.

This book pertains to the first strand of literature. The notion of outsourcing employed throughout this book refers to "general outsourcing" firstly coined by Chongvilaivan and Hur (2011) and thus takes into account vertical production linkages with both domestic and foreign outsourcing providers for two main reasons.

Firstly, and perhaps most importantly, manufacturing firms in Thailand pertain intimately to both international and domestic procurement of intermediate materials and services. They take advantage of the external outsourcing providers by contracting out their production locally and internationally to maintain their core activities and remain competitive in the global market. For example, the provision of IT services, including IT development and maintenance, has played an indispensable role of outsourcing activities in securing comparative advantages by reducing operational costs and obtaining better IT services from outside specialized providers. Outsourcing is therefore a strategic option for manufacturing firms in Thailand to lower their operation costs and maintain their competitiveness in the global market. In this regard, both domestic and international outsourcing activities have formed production fragmentation and vertical business linkages in Thailand.

In addition, the recent evidence has increasingly supported that it is general outsourcing, rather than international outsourcing, which is more relevant to economic impacts of production fragmentation. For instance, Chongvilaivan and Hur (2011) demonstrated that general outsourcing contributed more significantly to the observed labour productivity growth in the U.S. manufacturing industries. Geishecker and Görg (2008) likewise found that domestic outsourcing also exhibited a noticeable effect on labour productivity in Germany even though its effect seems to be less pronounced than that of international outsourcing. In the context of wage inequality between skilled and unskilled workers, Chongvilaivan et al. (2009) showed that general outsourcing was a crucial driver of wage inequality due to its skill-biased productivity effects, using industry-level information retrieved from the U.S. manufacturing data.

Following the definition of general outsourcing put forward by Abraham and Taylor (1996), Girma and Görg (2004), and Chongvilaivan and Hur (2011), the index of outsourcing (O_i^T) is measured by

$$O_i^T = \frac{\sum_j MOUT_{ij} + \sum_k SOUT_{ik}}{TC_i} \tag{1.1}$$

where $MOUT_{ij}$ denotes the uses of materials input j, including materials and component purchases, by an establishment i, and $SOUT_{ik}$ refers to the costs of services purchases k, including contract, commission, repair and maintenance work, by an establishment i, and TC_i represents total production costs by an establishment i.

Moreover, I also attempt to ponder the possibilities that different types of production activities taken by outsourcing providers may exhibit different impacts on factor productivity leading to different impacts on wage differentials (see Chapter 6). Accordingly, I distinguish the index of outsourcing (O_i^T) into two sub-categories: materials outsourcing (O_i^M) and services outsourcing (O_i^S).

$$O_i^M = \frac{\sum_j MOUT_{ij}}{TC_i} \text{ and } O_i^S = \frac{\sum_k SOUT_{ik}}{TC_i} \tag{1.2}$$

1.3 Overview of the Book

Having illuminated the motivation of this book, depicted a primer of production fragmentation in Thailand's manufacturing industries, and defined the notion of outsourcing this book refers to, I end the first chapter by enumerating the synopses of the remaining chapters. This section establishes itself as a helicopter tour that helps readers perceive the aims and scope of the ensuing chapters in this book. The rest of this book can be briefly outlined as follows.

My attempts in empirically investigating the economic effects of production fragmentation in Thailand's manufacturing industries begin with the examination of the characteristics of Thai manufacturing establishments. Chapter 2 is devoted to identifying the qualities an outsourcing firm — a firm that pertains to production fragmentation — has. This chapter traces out their various features, such as foreign ownership, exporting activities, and the use of information technology, among others. Following a thorough search for possible explanations, the results from the survey and analyses are presented.

Chapter 3 starts with raising the question: "Does production fragmentation entail higher productivity of firms?" This chapter proposes the mechanisms through which production fragmentation affects productivity of firms, and provides a thorough review of the literature. I then portray evidence based on the survey and analyse rationalizing the linkages.

Chapter 4 examines the relationship between production fragmentation and productivity of labour employed in-house. In this chapter, I first explore various hypotheses of the labour productivity effects based on existing research. These effects on labour markets are then analysed based on my Thailand's establishment-level dataset.

Central to the subject of Chapter 5 is skill upgrading — a shift in the relative demand for skilled labour. This empirical investigation essentially identifies winners and losers from production fragmentation. The evidence this chapter produces is useful in its own right because it provides intuitive insights on how production fragmentation explains worsened wage inequality between high- and low-skilled workers. This chapter is first motivated by theoretical propositions and then looks into the existing findings. Then, I turn to an examination of the establishment-level data.

Chapter 6 departs from the preceding chapters by distinguishing between materials and services fragmentation. The objective of this chapter is to examine the roles of fragmentation typology in explaining the economic effects analysed in Chapters 3–5. In this chapter, I expound the evolution of the recent debate on the roles differently played by materials and services fragmentation. Decomposing fragmentation into sub-categories may yield more refined economic consequences. My dataset is depicted and analysed thereafter.

Chapter 7 concludes this book. The last chapter, based on the discussions in Chapters 2–6, summarizes the policy implications and enumerates what kinds of industrial development policy, with growing prevalence of production fragmentation, help advance productivity of firms, promote skill development, and mitigate the negative effects of widening wage inequality.

2
Characteristics of Firms in Thailand's Manufacturing Sector

The objective of this chapter is to explore characteristics of the establishments operating in Thailand's manufacturing industries based on the establishment-level data retrieved from the reports of the *2003 Manufacturing Industry Survey*, provided by the National Statistical Office (NSO), Thailand. The main purpose of this survey is to collect basic information on industrial establishments with at least one person engaged throughout Thailand. This survey contains establishment-level information on production and cost structures such as numbers of employees, values of remuneration, raw materials, parts and components purchased, sales of goods produced, and fixed assets, among others. The information gathered contributed to the formulation of economic and industrial development policies as well as the decision-making process in both government and private sectors.

According to the survey, a manufacturing establishment refers to an establishment engaged in the mechanical or chemical transformation of substances into new products. The assembly of the component parts of manufactured products is also considered as manufacturing. Manufacturing activities are classified according to the 4-digit International Standard Industrial Classification (ISIC) Rev. 3.

The survey pertained to sixty-two categories of manufacturing activities (4-digit ISIC) in twenty-three industries (2-digit ISIC).

As portrayed in Table 2.1, there were 8,862 manufacturing establishments pertinent to this dataset. Food and beverages manufactures accounted for approximately 25 per cent of the total samples and thus were the largest in terms of the number of establishments, followed by furniture (9.4 per cent), fabricated metal products (8.81 per cent) and textiles (8.80 per cent). The large numbers of establishments in these industries may suggest that the establishments encountered the high degree of competition. In contrast, office, accounting and computing machinery, recycling, refined petroleum products and tobacco were relatively less competitive as there were merely 14, 33, 36 and 47 establishments, respectively, operating in these industries.

Table 2.2 reveals the international exposure of establishments in Thailand's manufacturing industries in terms of foreign ownership, exports and use of imported intermediate inputs. These attributes have crucial implications on outsourcing decisions now that establishments with intensive exposure to foreign markets tend to find vertical disintegration less costly than domestic establishments, thanks to their superior knowledge and/or information on foreign markets and input providers (Markusen 1995 and Girma and Görg 2004). The first column reports the numbers of foreign-owned establishments. The numbers of establishments that carried out exports is shown in the second column. The last column exposes the numbers of establishments which utilized imported intermediate input.

TABLE 2.1

Numbers and Percentages of Establishments by Manufacturing Activities

Industry	*No. of Est.*	*Per Cent*
Food and beverages	2,251	25.40
Tobacco	47	0.53
Textiles	780	8.80
Wearing apparel; dressing and dyeing of fur	337	3.80
Leather footwear products	278	3.14
Wood and cork, except furniture	333	3.76
Paper and paper products	187	2.11
Publishing, printing and reproduction of recorded media	232	2.62
Refined petroleum products	36	0.41
Chemicals and chemical products	374	4.22
Rubber and plastics products	440	4.97
Other non-metallic mineral products	723	8.16
Basic metals	143	1.61
Fabricated metal products	781	8.81
Machinery and equipment	313	3.53
Office, accounting and computing machinery	14	0.16
Electrical machinery and apparatus	141	1.59
Communication equipment and apparatus	141	1.59
Medical precision and optical instruments	96	1.08
Motor vehicles, trailers and semi-trailers	176	1.99
Other transport equipment	173	1.95
Furniture manufacturing	833	9.40
Recycling	33	0.37
Total manufacturing	**8,862**	**100**

TABLE 2.2

The Numbers and Percentages of Establishments with Foreign Ownership and Engaged in Export Activities and Imports of Intermediate Inputs

Industry	*Foreign Ownership*	*Exports*	*Use of Imported Input*
Food and beverages	118 (5.24)	309 (13.73)	211 (9.37)
Tobacco	2 (4.26)	3 (6.38)	n.a.
Textiles	62 (7.95)	121 (15.51)	101 (12.95)
Wearing apparel; dressing and dyeing of fur	19 (5.64)	64 (18.99)	46 (13.65)
Leather footwear products	29 (10.43)	64 (23.02)	72 (25.90)
Wood and cork, except furniture	12 (3.60)	67 (20.12)	63 (18.92)
Paper and paper products	27 (14.44)	46 (24.60)	50 (26.74)
Publishing, printing and reproduction of recorded media	10 (4.31)	8 (3.45)	27 (11.64)
Refined petroleum products	8 (22.22)	9 (25.00)	16 (44.44)
Chemicals and chemical products	70 (18.72)	80 (21.39)	127 (33.96)
Rubber and plastics products	82 (18.64)	147 (33.41)	122 (27.73)
Other non-metallic mineral products	45 (6.22)	99 (13.69)	84 (11.62)

Basic metals	19 (13.29)	28 (19.58)	34 (23.78)
Fabricated metal products	78 (9.99)	100 (12.80)	163 (20.87)
Machinery and equipment	55 (17.57)	72 (23.00)	105 (33.55)
Office, accounting and computing machinery	6 (42.86)	5 (35.71)	6 (42.86)
Electrical machinery and apparatus	41(29.08)	47 (33.33)	61 (43.26)
Communication equipment and apparatus	59 (41.84)	63 (44.68)	75 (53.19)
Medical precision and optical instruments	21 (21.88)	24 (25.00)	37 (38.54)
Motor vehicles, trailers and semi-trailers	48 (27.27)	48 (27.27)	65 (36.93)
Other transport equipment	26 (15.03)	25 (14.45)	40 (23.12)
Furniture manufacturing	97 (11.64)	219 (26.29)	172 (20.65)
Recycling	n.a.	n.a.	n.a.
Total manufacturing	**934 (10.54)**	**1,648 (18.60)**	**1,677 (18.92)**

Note: Percentages in parentheses.

In these columns, the percentages of the total numbers of establishments appear in parentheses. exposure to international markets in that foreign-owned and exporting establishments

Out of 8,862 samples, there were 934 foreign-owned establishments. Industrial decomposition further reveals that high-tech and capital-intensive manufacturing industries in Thailand tend to have relatively high foreign ownership, particularly office, accounting and computing machinery (42.86 per cent), communication equipment and apparatus (41.84 per cent), electrical machinery and apparatus (29.08 per cent), and motor vehicles, trailers and semi-trailers (27.27 per cent). Intensive foreign ownership pertaining to high-tech and capital-intensive production may imply that an industry with higher foreign ownership would have better access to cutting-edge technology. This proposition seems plausible given the observations that the figures are relatively low for labour-intensive industries including wood and cork (3.60 per cent), tobacco (4.26 per cent), publishing, printing and reproduction of recorded media (4.31 per cent), and food and beverages (5.24 per cent). These results are not unpredicted since in the context of developing Southeast Asia, high-tech production activities are concerned with cutting-edge production technology and know-how possessed by multinational enterprises (MNEs) — a key vehicle of technological spillovers.

Excluding publishing, printing and reproduction of recorded media and tobacco, Thailand's manufacturing industries are export-oriented with the proportion of exporting firms ranging from 12.80 per cent in the machinery and equipment industry to 44.68 per cent in the communication equipment and apparatus industry. The particularly large proportions of exporting establishments in some advanced industries, like communication equipment and apparatus (44.68 per cent), office, accounting and computing machinery (35.71 per cent) and electrical machinery and apparatus (33.33 per cent), suggest that Thailand's comparative advantages have departed from traditional, labour-intensive production, like food and beverages and textiles, to sophisticated, capital-intensive production. The structural adjustments in Thailand's manufacturing sector are attributable fundamentally to its outward-looking industrialization policy.

Establishments in Thailand's manufacturing sector also prevalently utilized imported intermediate inputs ranging from nearly 10 per cent for food and beverages to 53.19 per cent for communication equipment and apparatus. Again, the figures are strikingly high for advanced, capital-intensive industries, especially communication equipment and apparatus (53.19 per cent), refined petroleum products (44.44 per cent), electrical machinery and apparatus (43.26 per cent), and office, accounting and computing machinery (42.86 per cent). The widespread use of intermediate imports in Thailand's manufacturing sector was caused mainly by the emergence of production fragmentation that enabled an establishment to contract out its production activities at arm's length.

Having enumerated various attributes of the surveyed manufacturing establishments, I then explore their employment structures. Table 2.3 reports the numbers of total workers in the first column, the numbers of non-production workers in the second column, and the skill

TABLE 2.3

Numbers of Workers Engaged and Skilled Labour Intensity

Industry	*Total Workers*	*Non-Production Workers*	*Skill Intensity (%)*
Food and beverages	221,217	43,259	19.56
Tobacco	7,489	3,451	46.08
Textiles	85,827	8,496	9.90
Wearing apparel; dressing and dyeing of fur	41,406	3,898	9.41
Leather footwear products	42,205	6,530	15.47
Wood and cork, except furniture	26,980	2,909	10.78
Paper and paper products	18,633	3,190	17.12
Publishing, printing and reproduction of recorded media	12,381	3,414	27.57
Refined petroleum products	3,079	744	25.14
Chemicals and chemical products	32,818	8,276	25.22
Rubber and plastics products	67,412	8,944	13.27
Other non-metallic mineral products	57,105	9,420	16.50
Basic metals	10,795	1,706	15.80
Fabricated metal products	40,873	6,711	16.42
Machinery and equipment	35,297	5,926	16.79
Office, accounting and computing machinery	11,896	1,146	9.63
Electrical machinery and apparatus	24,709	3,554	14.38
Communication equipment and apparatus	56,568	5,951	10.52
Medical precision and optical instruments	14,023	1,160	8.27
Motor vehicles, trailers and semi-trailers	36,226	7,556	20.86
Other transport equipment	18,659	2,549	13.66
Furniture manufacturing	107,342	13,042	12.15
Recycling	438	88	20.09
Total manufacturing	**973,378**	**151,950**	**15.61**

Note: Skill intensity is measured by the ratio of non-production workers to total workers.

intensity, measured by the ratios of non-production to total workers, in the last column. As it will be discussed later in Chapter 5, the non-production workers in my dataset are pertinent to non-production activities, such as administrative, technical and clerical personnel such as managers, directors, laboratory and research workers, and administrative supervisors and the like, and hence are by definition more skilled than production workers which are deployed in the production or other related activities such as workers in factories, caretakers, machine tenders, machine controllers and assemblers.

Table 2.3 reveals that there were 973,378 workers employed in my sample manufacturing establishments, whilst food and beverages and furniture manufacturing industries seemed to be the largest in terms of employment. Furthermore, employment in Thailand's manufacturing sector was devoted largely to production activities since non-production workers accounted for approximately 16 per cent of the total employment. Except for the tobacco industry in which the proportion of non-production workers was strikingly high (46.08 per cent), skill intensity was dispersed in a tapered range of roughly 10–25 per cent. The relatively low non-production employment might be explained by the fact that Thailand's manufacturing sector was well-endowed with a large pool of cheap low-skilled workers who are usually employed in production activities. Therefore, firms would have an incentive to weigh in on production activities within Thailand's manufacturing sector.

Table 2.4 paints a clearer picture of human capital investment in Thailand's manufacturing industries and sheds

TABLE 2.4

Cost of Research and Development (R&D) and Labour Training as Percentages of Total Expense

Industry	*R&D*	*Training*
Food and beverages	1.01	0.31
Tobacco	n.a.	0.11
Textiles	0.62	0.21
Wearing apparel; dressing and dyeing of fur	1.27	0.76
Leather footwear products	4.92	0.35
Wood and cork, except furniture	0.05	0.15
Paper and paper products	0.28	0.29
Publishing, printing and reproduction of recorded media	0.29	0.25
Refined petroleum products	0.04	0.04
Chemicals and chemical products	0.41	0.25
Rubber and plastics products	0.92	0.34
Other non-metallic mineral products	0.12	0.44
Basic metals	0.07	0.20
Fabricated metal products	0.15	0.46
Machinery and equipment	4.28	0.91
Office, accounting and computing machinery	1.18	0.88
Electrical machinery and apparatus	0.58	1.12
Communication equipment and apparatus	0.28	0.33
Medical precision and optical instruments	11.20	0.80
Motor vehicles, trailers and semi-trailers	0.90	0.26
Other transport equipment	1.44	2.14
Furniture manufacturing	2.18	0.51
Recycling	n.a.	n.a.
Total manufacturing	**0.99**	**0.36**

further light on my previous observation that manufactures in Thailand were concerned predominantly with low skill-intensive production activities. In this table, costs of research and development (R&D) and labour training as percentages of the total expenses of an establishment are reported in the second and third columns, respectively.

The figures disclose a root cause of inadequate human capital accumulations in Thailand's manufacturing sector. On average, the sample manufacturing establishments devoted merely 1 and 0.4 per cent of their total operating expenses to R&D and training, respectively. At the industry level, the medical precision and optical instruments industry was the most outstanding in terms of R&D investment which accounted for approximately 11.2 per cent of the total expenses. Human capital under-investment characterizing Thailand's manufacturing sector was fundamentally attributable to its deficiency of the key drivers of a knowledge-based economy. These include skilled human resources as well as physical and institutional infrastructure.

A natural question arises: What is the key engine of Thailand's manufacturing sector growth? The answer to this question perhaps rests with the figures of information and communication technology (ICT) use reported in Table 2.5 which reveals the numbers of firms that made use of computers and internet in the second and third columns, respectively.

A key catalyst of proliferating outsourcing activities in Thailand rests with the rapid advancement of information and telecommunication technology (ICT) which allows companies to downsize its service function and

TABLE 2.5

The Numbers and Percentages of Establishments using Computers and Internet

Industry	*Computer*	*Internet*
Food and beverages	810 (35.98)	432 (19.19)
Tobacco	6 (12.77)	4 (8.51)
Textiles	258 (33.08)	151 (19.36)
Wearing apparel; dressing and dyeing of fur	106 (31.45)	67 (19.88)
Leather footwear products	115 (41.37)	82 (29.50)
Wood and cork, except furniture	137 (41.14)	65 (19.52)
Paper and paper products	131 (70.05)	78 (41.71)
Publishing, printing and reproduction of recorded media	192 (82.76)	85 (36.64)
Refined petroleum products	25 (69.44)	10 (27.78)
Chemicals and chemical products	219 (58.56)	151 (40.37)
Rubber and plastics products	304 (69.09)	207 (47.05)
Other non-metallic mineral products	333 (46.06)	173 (23.93)
Basic metals	70 (48.95)	42 (29.37)
Fabricated metal products	365 (46.37)	219 (28.04)
Machinery and equipment	202 (64.54)	131 (41.85)
Office, accounting and computing machinery	14 (100.0)	14 (100.0)
Electrical machinery and apparatus	91 (64.54)	66 (46.81)
Communication equipment and apparatus	109 (77.30)	92 (65.25)
Medical, precision and optical instruments	45 (46.88)	32 (33.33)
Motor vehicles, trailers and semi-trailers	119 (67.61)	78 (44.32)
Other transport equipment	72 (41.62)	48 (27.75)
Furniture manufacturing	345 (41.42)	242 (29.05)
Recycling	6 (18.18)	1 (3.03)
Total manufacturing	**4,074 (45.97)**	**2,470 (27.87)**

Note: Percentages in parentheses.

contract out some of these functions to lower cost locations. As shown in Table 2.5, establishments in Thailand's manufacturing industries prevalently brought into play information technology, in particular computers and internet. Approximately 46 per cent of the sample establishments employed computers in their production and management whereas the figure is about 28 per cent for the numbers of establishments that used internet services. The proportions of establishments utilizing computers and internet were strikingly high for the relatively advanced industries such as office, accounting and computing machinery, communication equipment and apparatus, electrical machinery and apparatus, and machinery and equipment.

With the use of ICT, the establishments were able to leverage on industrial development through logistics, R&D, information, digital content, design and cultural innovation. This evidence also highlights an increasingly important role of production fragmentation in Thailand's manufacturing industries, which necessitates the exploitation of ICT and serves as a tool for achieving industrial development, with limited human capital resources.

All in all, the examination of my establishment-level dataset points out several peculiar features of Thailand's manufacturing sector. Manufacturing industries have been the crucial engine of economic growth in Thailand since the past decades. They were highly open in terms of attracting FDI, exports and use of intermediate imports. This sector also absorbed a huge amount of workers; most of which, however, pertained to low skill-intensive production activities. Excessive concentration on such low value-added production activities was perhaps caused by human capital underinvestment — my analyses indicate that human capital investment in Thailand's manufacturing sector is nearly negligible. Inadequate human capital accumulations will ultimately constitute a major barrier of industrial development in Thailand. Interestingly, this chapter finds that the establishments were intensively utilizing ICT, especially computers and internet, notwithstanding their limited human capital resources. In this regard, the proliferation of the use of ICT in Thailand's manufacturing sector points to a pivotal role of production fragmentation as the key catalyst of industrial development and therefore calls for closer investigation of its impacts on the labour market. The discussions and empirical analyses developed later in this book provide several policy implications for the governments, especially those in developing Southeast Asia, who must step in and harness the impacts of production fragmentation to achieve sustainable economic growth and labour development.

3
Firm Productivity Effects

3.1 Introduction

Changes in productivity of a firm are perhaps the most fundamental question of the economic impacts of production fragmentation (or outsourcing). It appears to be widely accepted that outsourcing has become an increasingly imperative option for manufacturing and non-manufacturing sectors to lower their operation costs, enhance their profitability performance, and maintain their competitiveness in domestic and international markets. Görg and Hanley (2005) documented: "A recent survey involving CEOs from 440 product and service companies found that respondents who outsourced intermediates (material inputs and services), realised higher profits and revenue than those that did not." Productivity improvements outsourcing conveys emanate from specializing in in-house production of core-competent activities, accessing technical expertise unavailable in-house, and acquiring a larger, cheaper pool of intermediate materials and services.

The purpose of this chapter is to empirically examine the impacts of outsourcing on productivity performance by employing Thailand's establishment-level data. The simple empirical exercise presented in this chapter produces the following interesting findings. Firstly, outsourcing helps boost productivity of an establishment, measured by the ratio of output to capital assets in Thailand's manufacturing industries. Secondly, the productivity gains may be attributable to the fact that establishments with higher outsourcing intensity are more likely to employ high-tech capital, such as computer machinery and equipment, intensively, thereby augmenting its productivity. Thirdly, a breakdown of the establishment-level dataset into two categories — one which pertains to international outsourcing and the other which contracts out the production stages purely locally — reveals that productivity gains outsourcing entails tend to be more pronounced for the former. In this sense, outsourcing locations have a say in the relationship between outsourcing and productivity improvements. Last but not least, nationality of ownership matters. The empirical evidence shows that a foreign-owned establishment reaps greater benefits in terms of higher productivity from contracting out production stages at arm's length than does a locally-owned one.

This chapter contributes to the existing research that examines the impacts of outsourcing on productivity and performance at the firm or industry levels. Much literature has been devoted to empirically examining such a linkage in industrialized economies, especially the United States (U.S.), the European Union (EU), and many newly industrialised economies (NIEs).[1] The lack of evidence in developing countries is due largely to the absence of detailed statistics that enable one to analyse its economic consequences. The present chapter, to the best of my knowledge, is the first to empirically investigate the productivity effects of outsourcing in Thailand's manufacturing industries. Furthermore, the disaggregated manufacturing data at the establishment level illuminate clearer insight into how outsourcing alters productivity performance of an establishment.

The remainders of this chapter are organized as follows. Section 3.2 provides a brief review of the literature concerned with the productivity effects of outsourcing. Section 3.3 enumerates the mechanisms through which outsourcing enhances productivity of an establishment. Section 3.4 depicts the productivity measurement utilized in this paper. Section 3.5 presents and analyses the empirical evidence based on Thailand's establishment-level data. Section 3.6 concludes.

3.2 Review of the Literature

There are two strands of literature that investigate the productivity effects of fragmentation on manufacture productivity. On the one hand, a number of international trade literatures have researched on how imported intermediate inputs shape the productivity at firm and industry levels. On the other hand, recent research has increasingly argued that the domestic dimension of production fragmentation might also play an important role in explaining changes in firm performance. The former confines the notion of production fragmentation merely to its international components, namely *international outsourcing* (Feenstra and Hansen 1996*a* and 1996*b*), whereas the latter broadly contemplates both domestic and international components, namely *general outsourcing*

(Girma and Görg 2004; and Chongvilaivan and Hur 2011). Given these two different terminologies the existing studies ponder, this section is devoted to a brief review of both strands of literature.

Among the very first studies concerned with the first strand of literature — an analysis of international outsourcing and firm productivity — is that of Siegel and Griliches (1992). They empirically investigated the linkage between international outsourcing of services and total factor productivity (TFP) growth, using the U.S. manufacturing industry data. Their empirical results, however, manifest that the acceleration of TFP growth and services outsourcing are unrelated. The weak link may stem from data measurement errors for capital services and labour inputs.[2] Their results have spurred an explosion of more rigorous research on whether and how productivity growth of the U.S. manufacturing industries is attributable to services outsourcing. For instance, ten Raa and Wolff (2001) found that services outsourcing contributed to the TFP growth recovery during 1987–96. Productivity gains services outsourcing conveys can be explained by the fact that manufacturing industries, to an increasing extent, could contract out less efficient services activities and carry out their core-competent production activities in-house. Amiti and Wei (2009) analysed the effects of international outsourcing of materials and services on productivity growth in the U.S. manufacturing industries. Their findings indicate that international outsourcing of both materials and services significantly contributes to productivity growth, and are thus in contrast with those of Siegel and Griliches (1992). The breakdown of outsourcing indices reveals that international outsourcing of services substantiates productivity growth in the U.S. manufacturing industries.

One problem remains in those early studies. The use of aggregate data of manufacturing industries implies that the estimates fail to account for the large heterogeneity — such as the exposure to international trade and foreign ownership — embodied in the firm behaviour. Accordingly, several recent studies have prevalently focused on disaggregate establishment-level data, thereby yielding further insight into the impacts of production fragmentation and firm productivity.

In particular, Görg and Hanley (2005) analysed the effect of international outsourcing of services and materials inputs on firm productivity, using establishment-level data for Ireland's electronics industry. They found that international outsourcing in general has a positive, significant impact on the productivity of an establishment, measured by its value added. However, the breakdown of total international outsourcing reveals that its typology matters for the impacts on firm productivity improvements. Specifically, they find that international outsourcing of materials apparently brings about productivity improvements, but that of services does not. Moreover, productivity gains from international outsourcing of materials are dependent on the degree to which an establishment deals with exporting activities. Their estimates reveal that the establishments with low export intensity are most likely to reap benefits from procuring intermediate inputs from the international markets in that international outsourcing may provide them greater exposure to international markets, which helps streamline their production processes and optimize their capacity. In this sense, the establishments, which have already intensively pertained to exporting activities and thus have already enjoyed the benefits offered by greater exposure to international markets, are less likely to reap additional benefits from international procurement of intermediate inputs.

In contrast with Görg and Hanley (2005), Görg et al. (2008) took into account the endogeneity of input choices and hence employed TFP as a more appropriate measure of firm productivity.[3] They employed establishment-level data for Irish manufacturing industries and found rather distinct results that the positive effects of services outsourcing are more pronounced than those of materials outsourcing. Furthermore, they further showed that firm-level heterogeneity has a crucial role to play in the relationship between international outsourcing of services and productivity improvements — the positive, statistically significant productivity effects of services outsourcing hold only for exporters, regardless whether they are foreign-owned. A possible explanation of this empirical evidence is that exporters tend to face lower costs of procuring intermediate inputs and searching for potential foreign suppliers since they pertain to international production networks and thus possess superior information on and knowledge of where and how to procure intermediate inputs competitively.

Most studies pertinent to productivity impacts of production fragmentation or outsourcing have focused on the imports of intermediate materials and services, thereby confining the notion of outsourcing only to its international component. A number of recent studies, however, have argued that domestic outsourcing — the extent to which a firm locally contracts out production activities at arm's length and/or locally procures intermediate materials and services — may also have a similar role to play in explaining productivity improvements. Hence, it may be unappealing to underplay the productivity effects of domestic outsourcing. These studies which accounted for domestic outsourcing and loosely defined the notion of production fragmentation as general outsourcing constitute the second strand of literature.

One of the earliest attempts to estimate the effects of fragmentation that incorporates both domestic and international components is that of Girma and Görg (2004). They examined the impacts of general outsourcing, defined as "the cost of industrial services received", on TFP using the establishment-level data for the United Kingdom (UK) manufacturing industries.[4] Their empirical results reveal that the establishments which undertake outsourcing activities intensively tend to have high productivity growth. The positive, statistically significant productivity effects of outsourcing, in addition, are more pronounced for the foreign-owned establishments.

The positive relationship between general outsourcing and productivity growth is confirmed by Calabrese and Erbetta (2005). They observed a considerable increase in the number of establishments in the Italian automotive industry undertaking outsourcing activities during the last decade. Utilizing several measures of firm performance including growth, productivity, financial dependence and profitability, they found that the relationship between outsourcing and firm performance might not follow a linear tendency. Furthermore, the establishments with the high degree of outsourcing are characterized by a high rate of productivity growth.

In a nutshell, a review of literature elucidated in this section suggests that production fragmentation in terms of both domestic and international outsourcing, at least in the long run, tends to entail positive productivity effects at the industry and firm levels. However, its accurate productivity improvements are rather mixed and less clear-cut. One natural question arises: What is the mechanism through which production fragmentation affects productivity? This will be the subject of discussion in the next section.

3.3 Production Fragmentation and Productivity

Having surveyed the existing studies that examine the link between production fragmentation and firm productivity as an empirical question, we now view the hypothesis that production fragmentation affects productivity through a theoretical rationale lens. A production theory (Fuss et al. 1978) is where the theoretical background lies.

Assume that final goods production pertains to a multi-stage production process, ranging from the upstream production stages, such as research and development (R&D), product design, and intermediate materials and services procurement, to the downstream production stages, such as assembling, packaging, and marketing. Expressed in a primal form, a production function of a single gross output Q_i an establishment i, where $i = 1, \dots, n$, has the following expression.

$$Q_i = Q_i\ (K_i, L_i, O_i) \tag{3.1}$$

where K_i and L_i are given quantities of capital and labour, respectively and O_i is the index of production fragmentation or outsourcing. As analysed in Fuss et al. (1978), the real value-added function takes the following definition.

$$V_i(K_i, L_i, O_i) = O_i = O_i \tag{3.2}$$

One problem remains in the estimations of a primal production function (3.1). The choice of intermediate inputs employed in in-house production is endogenously chosen. The estimates corresponding to (3.1) fail to capture this endogeneity problem and hence may be neither unbiased nor consistent. Due to this reason, it is more appealing to estimate the value-added function (3.2) instead of the production function (3.1). Next step involves assuming a functional form of the value-added function (3.2). Following Egger and Egger (2006) and Chongvilaivan and Hur (2011), we assume the constant elasticities of substitution (CES) specification.[5]

$$V_i = e^{\alpha+\gamma O_i}\{(e^{\beta_K O_i}K_i)^\rho + (e^{\beta_L O_i}L_i)^\rho\}^{r/\rho} \tag{3.3}$$

The elasticities of substitution (σ), as commonly known, can be measured by $(1 - \rho)^{-1}$, and r captures the degree of scale economies. The productivity impacts of outsourcing, based on the CES value-added function (3.3), work through two channels: neutral and factor-augmenting technological changes. The former captured by γ portrays the mechanism through which outsourcing augments overall productivity of a firm — an increase in its value-added without distorting the capital-labour ratio.

In contrast, the latter, reflected by β_K and β_L for capital- and labour-augmenting effects respectively, accounts for the ensuing productivity gains each factor of production enjoys.

There are at least two possible channels through which outsourcing activities affect the productivity of a firm. Firstly and perhaps more importantly, outsourcing entails efficiency gains from specialization in terms of scale economies and distributing fixed costs. When firms contract out non-core production stages at arm's length, the existing factors of production employed in-house — in particular capital and labour — will be reallocated to and specialized in core-competent production stages. For example, General Motors (GM) may outsource the production design and parts and components and may internalize car production and assembly, whereas Apple may outsource the production of its iPod players and undertake R&D and product design in-house. Improved resources allocation and more concentrated production factors would expand output, push the production frontier outward and ultimately enhance total factor productivity.

The second type of productivity effects concerns the larger pool of cheaper and better intermediate inputs outsourcing provides to firms. Productivity improvements may result from better access to traded intermediate inputs, which may be available at lower costs and higher quality than those performed in-house. Paisittanand and Olson (2006) argued that outsourcing is often more efficient than undertaking all production stages internally because production costs are lower, and quality is higher with outsourcing. Hence, increasing use of outsourcing may bring about a direct boost in production productivity of a firm, thereby shifting its production frontier outward. Banks and financial institutions, for instance, can augment their efficiency by relocating some parts of multiple stages like Information Technology and Human Resources departments, in which they are relatively inefficient, to other firms which are able to carry out those activities at lower costs.

3.4 Establishment-specific Productivity Measurement

As expounded in Section 3.2, a number of productivity measurements have been employed in the literature that examines the relationship between production fragmentation and its productivity effects. Among these measures, TFP growth is the most appropriate and the most widely adopted due to the following reasons.[6] First, the use of TFP growth accounts for the contemporaneous levels of technology that affect the optimal choices of production inputs (Levinson and Petrin 2003). This implies that production factors should be deemed as potentially endogenous in the estimation. More importantly, the relationship between production fragmentation and productivity may be endogenous if there are unobserved variables that are correlated with the decision to carry out production fragmentation and error terms. Phrased in another way, the establishment-specific productivity levels may influence a decision whether a firm will engage in production fragmentation — a more productive firm may be more likely to undertake production fragmentation, for instance.

Nevertheless, the TFP growth as a measure of establishment-specific productivity levels is not feasible in the context of Thailand's establishment-level dataset since the nature of the establishment-level data retrieved from the *2003 Manufacturing Industry Survey* is cross-sectional. Accordingly, the alternative, perhaps second-best, measure of establishment-specific productivity levels will be the output values scaled by capital assets, y_k, as follows.

$$y_{ki} = \frac{Y_i}{K_i} \tag{3.4}$$

where Y_i is total sales of goods produced by an establishment i, and K_i is the operating capital assets of an establishment i, including land, buildings, construction, machinery and equipment. The ratio of total sales to capital assets (y_k) as a proxy of the establishment-specific productivity levels requires the assumption that the part of the error term that is correlated with the choices of factor inputs or outsourcing variables is time-invariant. Although this assumption is by all means too strong, this can be the case since the interpretation of the parameter estimates based on the cross-sectional dataset has to do with the productivity effects of production fragmentation at a particular point of time. The empirical results, therefore, should be interpreted as instantaneous productivity effects of outsourcing.

3.5 Empirical Evidence from Thailand's Establishment-level Data

This section discusses and analyses the relationship between outsourcing (production fragmentation) and firm productivity, on the ground of empirical evidence from Thailand's establishment-level data. It should be highlighted that the parameter estimates obtained below are based on the full establishment-level sample size which amounts to more than 6,000 establishments, whereas the figures, for the purpose of presentation simplicity, are represented by aggregating establishments at their four-digit ISIC manufacturing industries, yielding sixty-three industries.[7]

As shown in Figure 3.1, the ratio of output to capital assets is positively correlated with the index of outsourcing (O^T) and statistically significant at 1 per cent.[8] This implies that the establishments which pertain to outsourcing activities tend to have higher productivity in terms of output per capital assets. Productivity gains from outsourcing are particularly consistent with a number of previous studies, including ten Raa and Wolff (2001), Görg and Hanley (2005), Görg et al. (2008) and Amiti and Wei (2009), among many others. The empirical results presented in this chapter are complementary to the literature on the impacts of outsourcing on firm productivity, which employs the data retrieved mostly from industrialized countries. The findings based on Thailand's establishment-level data further reveal that productivity gains from outsourcing, which allows the establishments to zero in on core-competent activities and to utilize cheaper and better intermediate inputs procured

FIGURE 3.1
A Fitted Plot of Outsourcing Intensity and Output-to-Capital Ratio

$$y_{ki} = .1696 + 3.1787^{***}O_i^T + u_i$$
$$(.1768) \quad (.2074)$$
$$N = 6602 \; ; \; F = 234.83^{***} \; ; \; R^2 = .0344$$

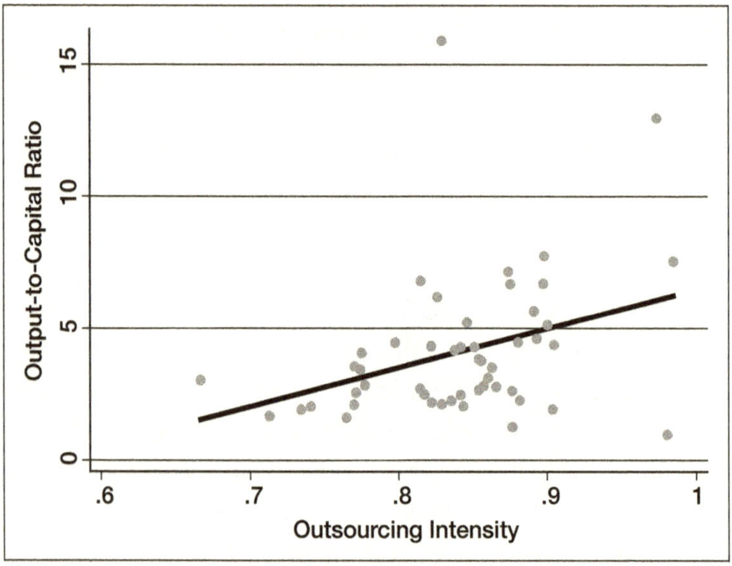

Notes: (1) *** represents statistical significance at 1 per cent; and (2) standard errors reported in parentheses.

from the markets, are also observed in a developing, less skill-intensive country like Thailand.

Figure 3.2 provides a robustness check of and sheds further light on the earlier result. In this figure, productivity of an establishment is replaced by the ratio of high-tech capital to capital assets. The high-tech capital includes computer machinery and equipment. The use of the high-

FIGURE 3.2

A Fitted Plot of Outsourcing Intensity and High-Tech Capital Intensity

$$T_i = .2619^{***} + .0323^{**}O_i^T + u_i$$
$$\quad (.0122) \qquad (.0143)$$
$$N = 6790 \; ; \; F = 5.09^{**} \; ; \; R^2 = .0007$$

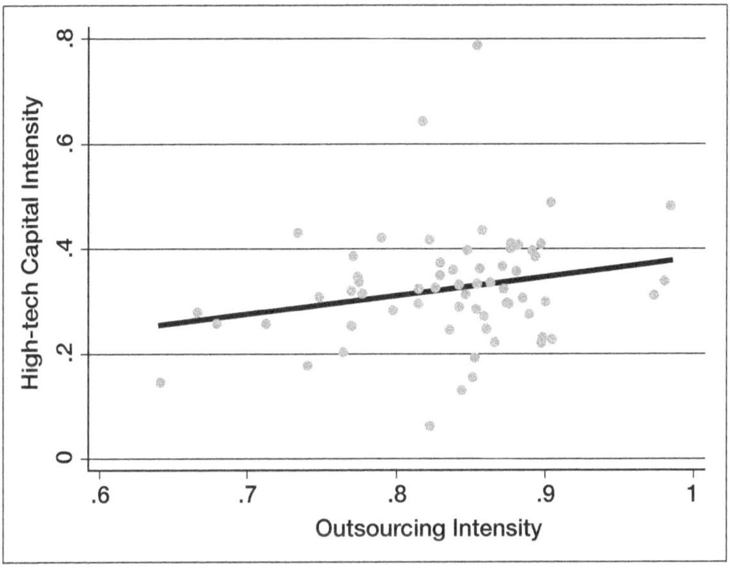

Notes: (1) ** represents statistical significance at 5 per cent; and (2) standard errors reported in parentheses.

tech capital intensity as a substitute of the output-capital ratio can be rationalized by the findings by Siegel and Griliches (1992) and Siegel (1997) who showed that high-tech capital formation is one of vital sources of productivity growth — a high-tech capital accumulation brings about higher productivity performance, thereby enhancing productivity growth and profitability.

The empirical estimates confirm a positive relationship between outsourcing and high-tech capital intensity since the coefficient of O^T (equal to .0323) is positive and statistically significant at 5 per cent. This indicates that establishments that carry out production fragmentation and outsourcing tend to employ high-tech capital intensively. In this sense, one would expect higher productivity growth from high-tech capital accumulations enhanced by outsourcing activities.

As depicted in Section 3.2, much research has been devoted to how an international outsourcing establishment — one which employs imported intermediate inputs — augments its productivity even though a number of recent studies have increasingly taken into account the role of domestic outsourcing which may have the similar role as international outsourcing to play in shifting an establishment's productivity levels and growth. Given the importance of outsourcing locations examined in the existing literature, it may be indispensable to reveal different productivity effects of international and domestic outsourcing in the context of Thailand's manufacturing industries.

In so doing, I, using the dummy of imported intermediate inputs, divide the establishment-level dataset into two groups. The first group, namely international outsourcing establishments, pertains to the establishments which engage in international outsourcing. The other, called domestic outsourcing establishments henceforth, includes those which procure intermediate materials and services merely locally. It should be emphasized that an international outsourcing establishment may also procure intermediate materials and services from domestic sources.

FIGURE 3.3

Fitted Plots of Outsourcing Intensity and Output-to-Capital Ratio among International Outsourcing Establishments

$$y_{ki} = .3072 + 4.6855^{***}O_i^T + u_i$$
$$(.5684) \quad (.6382)$$
$$N = 1421 \; ; \; F = 53.91^{***} \; ; \; R^2 = .0366$$

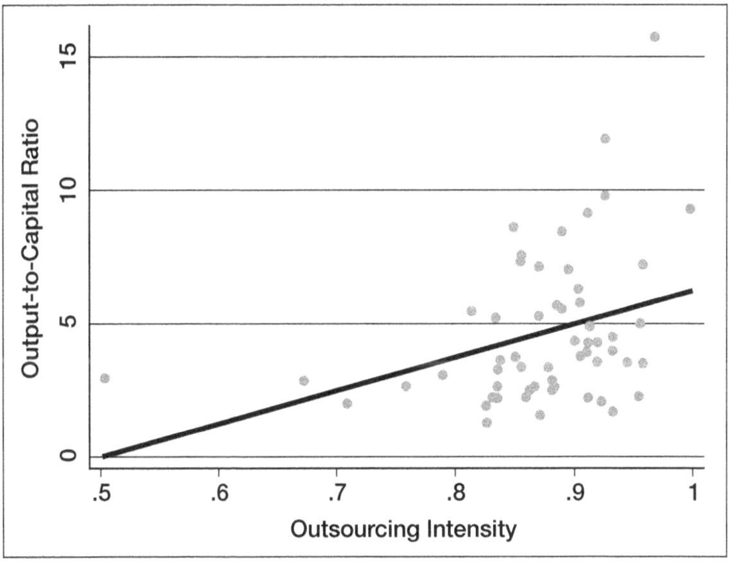

Notes: (1) *** represents statistical significance at 1 per cent; and (2) standard errors reported in parentheses.

Furthermore, it should also be acknowledged that the lack of intermediate imports data rules out the possibility to divide the activities of an international outsourcing establishment into international and domestic outsourcing.

The empirical evidence based on Thailand's establishment-level data is portrayed in Figures 3.3 and 3.4. I find the following interesting results. First, the relationship

FIGURE 3.4

Fitted Plots of Outsourcing Intensity and Output-to-Capital Ratio among Domestic Outsourcing Establishments

$$y_{ki} = .3388 + 2.6838^{***}O_i^T + u_i$$
$$(.1814) \quad (.2156)$$
$$N = 5181 \ ; \ F = 154.98^{***} \ ; \ R^2 = .0291$$

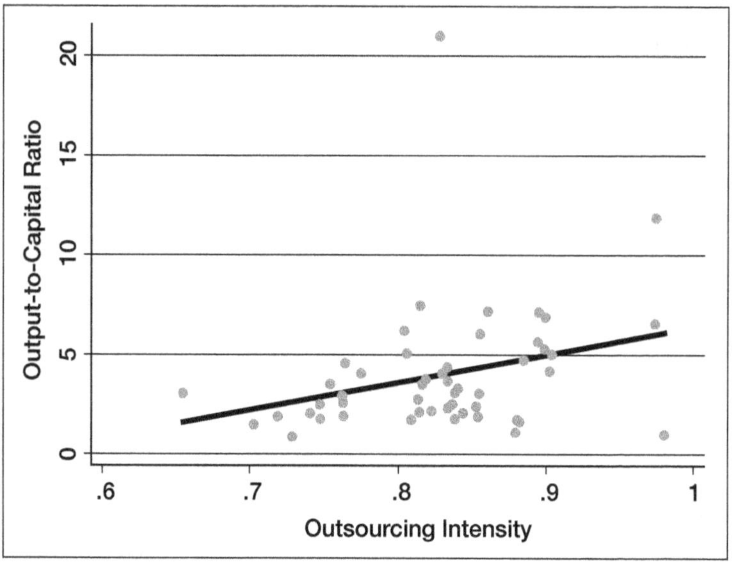

Notes: (1) *** represents statistical significance at 1 per cent; and (2) standard errors reported in parentheses.

between the ratio of output to capital assets and the index of outsourcing intensity is positive and statistically significant at 1 per cent for both types of establishments. Consistent with the previous results, Figure 3.3 implies that an establishment that carries out international outsourcing tends to enjoy higher productivity in terms of output per unit of capital assets. More importantly, the productivity effects of outsourcing seem to be more pronounced for an international outsourcing establishment since the coefficient of O^T for this grouping is greater than that of the domestic outsourcing firms (4.6855 > 2.6838). The evidence that productivity effects of an international outsourcing establishment are more magnified than those of a domestic outsourcing one implies that international outsourcing enhances productivity of an establishment. More interestingly, domestic outsourcing — local procurement of intermediate inputs — also brings about productivity improvements. This empirical exercise therefore substantiates the pivotal role of domestic outsourcing that explains increases in productivity of Thailand's manufacturing industries and hence goes along with the existing findings by Girma and Görg (2004).

Several studies also postulate that the nationality of ownership of an establishment may matter for the productivity impacts of outsourcing. One would expect foreign-owned establishments to have more drastic productivity effects of outsourcing for the following reasons. First, a foreign-owned establishment is by definition part of a multinational production network and is therefore expected to have better access to firm-specific assets and higher levels of technology (Markusen 1995). The intensive use of firm-specific assets and high technology will in turn intensify the productivity impacts of contracting out stages of production at arm's length. The second rationale is closely related to the first point. A foreign-owned establishment deals with the multinational vertical structure of production and is thus more likely to benefit from productivity gains from specialization than a locally-owned one. In this sense, outsourcing that allows an establishment to contract out non-core activities and specialize in core-competent activities, tends to be more advantageous for a foreign-owned establishment. Last but not least, since a foreign-owned establishment is by nature embodied in an international production network through its vertical linkages with the parent and other affiliates, it is expected to have better, perhaps more subtle, strategies for carrying out outsourcing activities and business connections with a pool of efficient providers than a locally-owned establishment.

To capture the role of ownership nationality, the dataset is again separated into two groups. The first is concerned with the foreign-owned establishments which by definition have at least partial foreign ownership, and the other includes the locally-owned establishment whose ownership is purely domestic.

Figures 3.5 and 3.6 manifest the fitted plots of the outsourcing intensity and the output-to-capital ratio for those two groups. These figures reveal the following interesting results. Firstly, outsourcing, as is expected, contributes to productivity improvements for both foreign- and locally-owned establishments. Moreover, the fitted plots show that foreign-owned establishments engage in outsourcing activities more intensively than do locally-owned establishments. As highlighted previously, this may be explained by the fact that a foreign-owned establishment is by definition concerned with a vertical international network and thus more likely to undertake outsourcing due to better information of intermediates providers and

FIGURE 3.5

Fitted Plots of Outsourcing Intensity and Output-to-Capital Ratio among Foreign-Owned Establishments

$$y_{ki} = -.8950 + 4.9532^{***}O_i^T + u_i$$
$$(.7265) \quad (.8154)$$
$$N = 811 \; ; \; F = 36.90^{***} \; ; \; R^2 = .0436$$

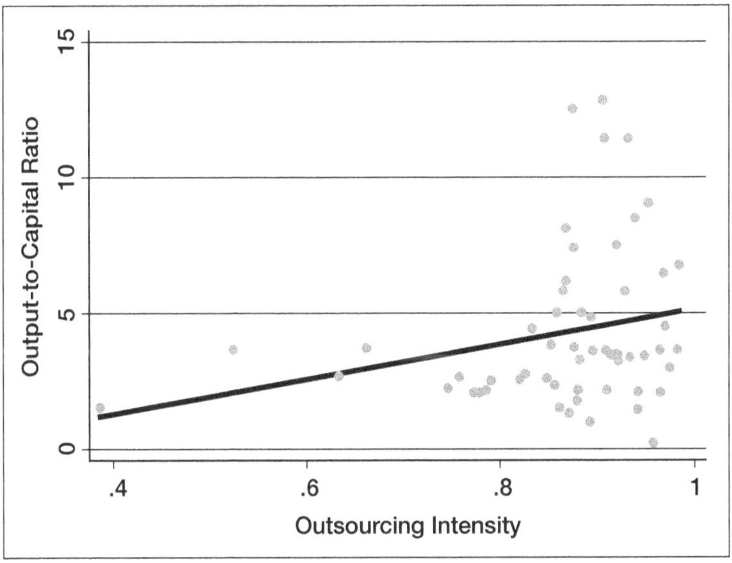

Notes: (1) *** represents statistical significance at 1 per cent; and (2) standard errors reported in parentheses.

more efficient contracting out strategies. Lastly and most importantly, the nationality of ownership matters to the productivity effects of outsourcing. The figures indicate that the impacts of outsourcing on productivity, though positive and statistically significant at 1 per cent in all cases, are more pronounced for foreign-owned establishments

FIGURE 3.6
Fitted Plots of Outsourcing Intensity and Output-to-Capital Ratio among Locally-Owned Establishments

$$y_{ki} = .2670 + 2.9717^{***}O_i^T + u_i$$
$$(.2147) \quad (.1818)$$
$$N = 5791 \; ; \; F = 191.58^{***} \; ; \; R^2 = .0320$$

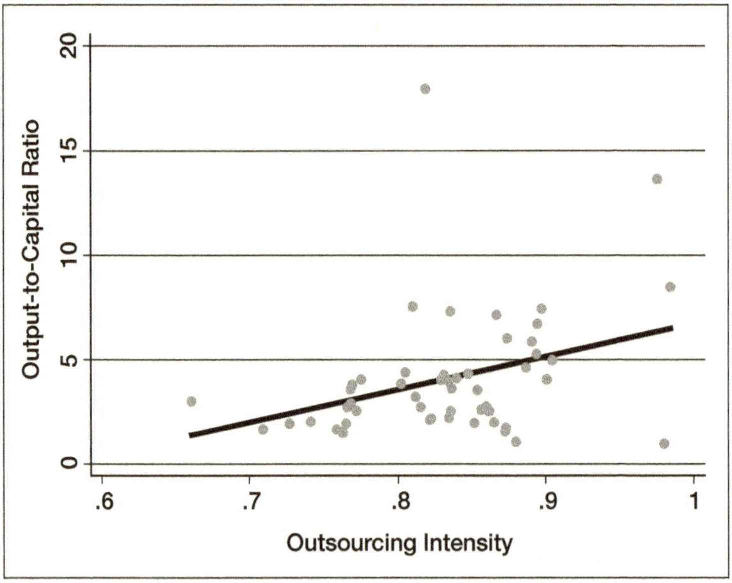

Notes: (1) *** represents statistical significance at 1 per cent; and (2) standard errors reported in parentheses.

(4.9532 > 2.9717). This result is strikingly consistent with that of Girma and Görg (2004), who found that the productivity gains from outsourcing activities are more prominent for foreign-owned establishments in the UK chemical and engineering sectors.

3.6 Concluding Remarks

This chapter has presented a simple empirical analysis of productivity effects of outsourcing using Thailand's establishment-level data. The productivity level is loosely defined as the output per unit of capital assets. The empirical evidence asserts that outsourcing is a crucial driver of productivity improvements in Thailand's manufacturing industries and hence is particularly consistent with a number of existing studies, which are primarily concerned with outsourcing in industrialized countries. The mechanisms through which contracting out fragmented stages of production brings about productivity shifts may be explained by gains from specialization. Particularly, outsourcing enables an establishment to contract out non-core activities at arm's length and specialize in in-house production of core-competent activities, thereby augmenting its overall productivity levels.

This chapter also attempts to account for establishment heterogeneity — locations of outsourcing and nationality of ownership — that may also make up the productivity effects of outsourcing. The findings reveal that the productivity shifts are more pronounced for the establishments which carry out international outsourcing. The more magnified productivity gains an international outsourcing establishment enjoys may be attributable to the fact that international outsourcing provides better access to more efficient providers operating overseas and a greater variety of intermediate inputs. A positive relationship between outsourcing and productivity levels characterizing the establishments which contract out production stages only locally sheds further light on the crucial role of domestic outsourcing in bolstering its productivity.

Dividing the dataset into foreign- and locally-owned establishments further reveals that the nationality of an establishment's ownership matters. A foreign-owned establishment tends to tap greater benefits in terms of productivity shifts from outsourcing than does a locally-owned establishment. This result is not surprising, however, since foreign-owned establishments have several advantages over those locally-owned. These include more available firm-specific assets, higher levels of technology, better outsourcing strategies, and a larger pool of potential providers.

Notes

1. See Section 3.2 for a brief review of literature.
2. Siegel and Griliches (1992) measure capital services and labour inputs by capital stocks and employment headcounts, respectively.
3. Using value added of an establishment as a measure of its productivity does not address the simultaneity problem in production function estimations, thereby potentially conveying inconsistent estimates.
4. The definition of outsourcing employed in Girma and Görg (2004) is analogous to that of Abraham and Taylor (1996). Abraham and Taylor defined outsourcing activities as various activities, including contracting out of machine maintenance services, engineering and drafting services, accounting services, computer services and janitorial services.
5. Several studies, such as Girma and Görg (2004) and Görg et al. (2008), among many others, employed the Cobb-Douglas production function for an empirical analysis of the relationship

between outsourcing and productivity. However, the Cobb-Douglas functional form assumes unit elasticities of substitution and is thus less flexible than the CES one.

6. A number of the existing studies, including Siegel and Griliches (1992), ten Raa and Wolff (2001), and Görg et al. (2008), among many others, employed TFP growth as a measure of firm productivity.
7. The reason for so doing is that the sample size is remarkably tremendous. As a result, it is difficult for one to capture any patterns from the fitted plot based on the establishment-level data.
8. The estimate equal to 3.1787 is positive and statistically significant at 1 per cent.

4
Labour Productivity Effects

4.1 Introduction

The past few decades witnessed a remarkable increase in production fragmentation and outsourcing activities in the world (Abraham and Taylor 1996; and Díaz-Mora 2007). The dominance of outsourcing which has become part and parcel of business practices is attributed to persistent plunges in transaction costs as a consequence of the prevalent uses of the information and transportation technology and trimmed trade barriers. The lack of intuitive understandings of its impacts on domestic labour markets gives rise to an exploding number of studies that inquire one of the main research questions in the literature — how does outsourcing elicit changes in labour productivity? The production theory suggests that outsourcing may be beneficial for productivity of labour employed in-house in at least two ways (Amiti and Wei 2009). First, the decision to outsource intermediate inputs is concerned with compositional changes. Fragmented stages of production allow a firm to contract out non-core production activities and reallocate the existing labour to core-competent activities. As workers become more concentrated on and/or specialized in the activities in which a firm possesses comparative advantage, one would expect higher productivity and performance workers can exert. Furthermore, outsourcing also results in structural changes — an outsourcing firm tends to have advantages over an integrated firm in terms of the superior access to a cheaper pool, more variety and higher quality of intermediate inputs. These structural improvements ultimately enable labour to materialize higher productivity via learning.

Using Thailand's establishment-level data, the present chapter empirically investigates the impacts of outsourcing on labour productivity, measured by output per worker. My empirical estimates reveal that outsourcing contributes to labour productivity growth in Thailand's manufacturing industries and hence are rather consistent with the findings by previous research, including Görzig and Stephan (2002), Egger and Egger (2006), and Amiti and Wei (2009), among many others. I also attempt to account for two establishment-specific characteristics — outsourcing locations and foreign ownership. The extension confirms the findings by Chongvilaivan and Hur (2011) that the effects of outsourcing on labour productivity are more pronounced for an establishment that pertains to international outsourcing, and therefore sheds further light on the crucial roles of outsourcing locations in explaining a shift in labour productivity. Furthermore, a breakdown of my establishment-level samples into foreign- and locally-owned establishments yields the results contrary to Girma and Görg (2004) that foreign ownership may have a limited role to play in the relationship between outsourcing activities and labour productivity improvements.

The main contribution lies with the fact that the present chapter, to the best of my knowledge, is among the very first attempts to empirically examine the impacts of production fragmentation and outsourcing on Thailand's labour market with a focus on changes in labour productivity.[1] It should be highlighted that this chapter approaches this research subject in a much simpler way than does the existing literature but illuminates clearer insights into the productivity effects of outsourcing on labour performance. The empirical results presented here therefore call for more detailed, in-depth future research on this subject.

The rest of this chapter is briefly organized as follows. Section 4.2 comprehensively reviews the existing studies that explored a linkage between outsourcing and labour productivity growth. Section 4.3 discusses a theoretical concept of the effects of outsourcing on labour productivity employing a Constant Elasticity of Substitution (CES) production function framework. Section 4.4 depicts labour productivity measurements. Section 4.5 presents and analyses the empirical results, based on Thailand's establishment-level data. Section 4.6 enumerates concluding remarks.

4.2 Review of the Literature

In recent years, one of the core research interests in the economic impacts of production fragmentation has been predominantly concerned with a change in labour productivity. Even though a prevalent use of production fragmentation has been widely seen as a crucial source of observed labour productivity growth, a theoretical discussion that addresses the impacts on labour productivity remains rather ambiguous, depending on whether labour is substituted by or complementary to the contracted out production activities, which sectors pertain to production fragmentation, factor and product market imperfections, and the intersectoral and international labour mobility.[2]

Among the very first attempts to empirically examine the effects of contracting out stages of production at arm's length on labour productivity is that of Görzig and Stephan (2002). They investigated how outsourcing activities prompt changes in labour productivity (measured by returns per employee), using firm-level panel data from the German cost structure survey over the period from 1992 to 2000. Three measurements of outsourcing activities they considered include intermediate material inputs, farming out production, and external services (e.g. consultancy or auditing). These measures are analogous to that of Abraham and Taylor (1996) and thus capture both domestic and international components. Their findings indicated that all three types of outsourcing activities entail labour productivity improvements; however, only intermediate material inputs are cost-saving — the other two types are not. These findings may suggest that firms have overestimated the benefits accruing from these two latter types of outsourcing, and thus outsourcing activities that successfully enhance labour productivities lie with well-functioning markets for intermediate inputs.

A later study by Girma and Görg (2004) investigated labour productivity gains in terms of changes in output per employee from outsourcing, employing establishment-level data for the United Kingdom (UK) manufacturing industries. Following Görzig and Stephan (2002), the notion of outsourcing in this study is loosely defined as "the contracting out of activities that were previously performed within a firm, to subcontractors outside the firm", including contracting out of machine maintenance services and engineering and drafting services. Girma and Görg showed that the labour productivity improvements might not hold at the industry-specific level. That is, labour employed in the chemical and engineering industries tend to enjoy its productivity gains whereas the influence of outsourcing on labour productivity changes is insignificant in the electronics industry. Their estimates further reveal that the labour productivity shifts are more pronounced for the foreign-owned establishments.

In contrast to Görzig and Stephan (2002) and Girma and Görg (2004), Egger and Egger (2006) offered a more detailed analysis of international outsourcing and its effects on productivity of low-skilled workers using data on twenty-two manufacturing industries in twelve European Union (EU) nations during 1992–97. Using a narrow definition of international outsourcing (see Feenstra and Hanson 1999) and the framework of the Constant Elasticity of Substitution (CES) production function, they found that international outsourcing undermines productivity of low-skilled labour in the short run — specifically, a one per cent increase in the international outsourcing intensity implies an approximately 0.2 per cent drop in real value added per low-skilled worker. In the long run, nevertheless, their productivity improvements stemming from international outsourcing can be observed. The short-run productivity deterioration and long-run productivity improvement associated with international outsourcing may be explained by imperfections in the European labour and goods markets, which hamper an immediate adjustment of the factors employment and the production structure.

Findings by Amiti and Wei (2009) shed further light on an insight into the impacts of international outsourcing on labour productivity. They explicitly distinguished international outsourcing of services from that of materials, using industry-level data of the United States (U.S.) manufacturing industries between 1992 and 2000. Their estimates reveal that services outsourcing has a larger role to play in shifting labour productivity than does materials outsourcing. Specifically, services outsourcing accounts for 10 per cent of the average growth in value added per worker while the contribution of materials outsourcing is much weaker — merely 5 per cent. These results imply that international outsourcing of services, relative to materials imports, is a key driver of labour productivity growth in the U.S.

Chongvilaivan and Hur (2011) viewed the linkages between outsourcing and its subsequent effects on labour productivity from a different angle. Building upon the CES framework as in Egger and Egger (2006) and using the six-digit NAICS U.S. manufacturing industries from 2002 to 2005, this study attempts to bridge the gap in the literature by employing two definitions of outsourcing: general outsourcing and international outsourcing. The former follows that of Abraham and Taylor (1996) and hence takes into consideration intermediate inputs procured from both local and international sources whereas the latter confines the notion of outsourcing only to intermediate imports as in Feenstra and Hanson (1996*a* and 1996*b*). They find that both notions of outsourcing play a similar role

— they augment productivity of both skilled and unskilled workers but are skill-biased. Their estimates obtained by estimating a primal CES production function further reveal that the productivity improvements skilled and unskilled workers have are more pronounced for general outsourcing. These results imply that contracting out production stages benefits labour employed in-house, regardless of the locations they are procured or undertaken at arm's length.

A growing number of studies attempted to examine the relationship between the more prevalent use of imported intermediate inputs and labour productivity shifts in the context of developing countries. For instance, Amiti and Konings (2007), rather than measuring the international outsourcing intensity as in Feenstra and Hanson (1996*a*, 1996*b* and 1999), employed an alternative measure of international outsourcing — input tariffs — to produce important new findings that real value added per worker as one of productivity measures is negatively correlated with the input tariffs, using establishment-level data retrieved from the Indonesian manufacturing census from 1991 to 2001. They argued that a higher usage of cheaper imported intermediate inputs contributes to labour productivity improvements in three ways: (1) learning effects from the foreign technology embodied in the imported intermediates; (2) quality effects from higher-quality inputs; and (3) variety effects from more input varieties. Furthermore, Thangavelu and Chongvilaivan (2010) employed the wide definition of international outsourcing as in Feenstra and Hanson (1996*b*) to empirically investigate the effects of outsourcing on labour markets using Thailand's manufacturing dataset from 1999 to 2003. Their parameter estimates indicate that skilled workers tend to reap benefits from international outsourcing of services while the effects on unskilled workers are insignificant. In contrast, international outsourcing of materials deteriorates productivity of unskilled workers whereas the skilled are unaffected. Their findings therefore are consistent with those of Amiti and Wei (2009) — international outsourcing of services, as opposed to that of materials, has emerged as a key driver of labour productivity growth.

To conclude, a brief review of the literature discussed thus far paints a clearer picture of how production fragmentation accounts for labour productivity growth. Much research has asserted rather consistent findings that contracting out stages of production in general contributes to a rise in productivity of workers employed in-house. Nevertheless, the more detailed effects on labour productivity, like those of firm productivity (see Chapter 3), counts crucially on where intermediate inputs are procured (e.g. international or domestic outsourcing); where the production plants are located (e.g. developed or developing countries); which types of production activities (e.g. materials or services) are contracted out, among others.

4.3 Production Fragmentation and Labour Productivity

A discussion on the linkage between production fragmentation and labour productivity developed in this section is based on the CES production framework, which have been introduced in Chapter 3, as in Chongvilaivan and Hur (2011). In principle, how outsourcing affects productivity of labour is captured by its marginal value added — how an incremental unit of labour augments a firm's value addedness.[3] Value added of labour, denoted by MV_{Li}, can be straightforwardly obtained by differentiating the CES production function (3.1) and taking a natural logarithm.

$$\ln MV_{Li} = \ln(rL_i^{\rho-1}) + \rho\beta_L O_i + \frac{\rho}{r}(\alpha + \gamma O_i) + \left(1 - \frac{\rho}{r}\right)\ln V_i \qquad (4.1)$$

where $MV_{Li} = \partial V_i / \partial L_i$

The effects of outsourcing on labour productivity can be represented by the elasticities of labour productivity (MV_{Li}) with respect to labour employed, denoted by η_{LOi}. Therefore, the next step is to differentiate Equation (4.1) with respect to $\ln O_i$.

$$\eta_{LOi} = \rho\beta_L O_i + \frac{\gamma\rho}{r} O_i + \left(1 - \frac{\rho}{r}\right)\eta_{VOi} \qquad (4.2)$$

where $\eta_{LOi} = \partial \ln MV_{Li} / \partial \ln O_i$, and $\eta_{VOi} = \partial \ln V_i / \partial \ln O_i$ is the elasticities of value added with respect to labour.

The primal approach to productivity analysis, based on the CES production function, is in contrast with the existing literature which conventionally makes use of a log-linearized Cobb-Douglas production function. It enables one to reveal the labour productivity effects in more details. The expression (4.2) comprising three adding-up terms breaks down total effects of outsourcing on labour productivity into three components: *factor-productivity effect*, *technology effect* and *value-added effect*.

The first term captures the factor-productivity effect of outsourcing — the partial improvements of labour productivity *vis-à-vis* labour-augmenting technological shifts. It essentially takes into account how outsourcing directly augments labour productivity. One could also easily show that the total effects on capital are the same as those on labour as portrayed in Equation (4.2), except for the first term in which β_L is replaced by β_K. The second term, the technology effect, represents labour productivity improvements triggered by neutral technological progress — technology improvements that uniformly augment productivity of all production factors. The last term, namely the value-added effect, is meant to account for the impacts of outsourcing on overall value added. It should also be observed that the absence of labour-augmenting effects of outsourcing ($\beta_L = 0$) does not mean labour fails to reap benefit from higher productivity boosted by outsourcing activities since it can still enjoy productivity improvements in terms of a surge in technological levels and overall value added.

4.4 Labour Productivity Measurement

There are at least three measures of labour productivity employed in the literature dealing with production fragmentation and its effects on labour markets — marginal productivity of labour, output per worker and value added per worker.

Marginal productivity of labour is perhaps the most meaningful, at least theoretically, among three measures of labour productivity since it captures the impacts of outsourcing on productivity of an additional worker employed. However, this measure is unobservable in the dataset. Chongvilaivan and Hur (2011) tackled this problem by utilizing a primal approach to a productivity analysis, in which marginal productivity of labour can be obtained by estimating a primal production function. The use of marginal productivity of labour as a proxy of labour productivity, though by all means appealing, requires the rigorous econometric modelling and discussions and is hence outside the scope of this book.

Alternative to marginal labour productivity is the value added per labour defined as the ratio of value added (or returns) to the number of labour employed. This measure of labour productivity is commonly employed in various studies such as Amiti and Wei (2009), Amiti and Konings (2007), Egger and Egger (2006), and Görzig and Stephan (2002) since it possesses several advantages. First, the measurement of value added per employee is straightforward and can be retrieved directly from most industry- and establishment-level datasets. More importantly, it also accounts for the endogeneity of intermediate inputs (e.g. intermediate materials and services); failing which may result in biased and inconsistent estimates of the impacts on labour productivity. Using value added per worker, nevertheless, is infeasible in the context of Thailand's manufacturing industries since the data of an establishment's value added are not available in the dataset retrieved from *The 2003 Manufacturing Industry Survey*.

The measure of labour productivity this chapter employs, therefore, is based on output per worker, defined as the ratio of an establishment's sales to the number of workers, as in Girma and Görg (2004).

$$y_{li} = \frac{Y_i}{L_i} \tag{4.3}$$

Even though this measurement is rather uncomplicated, it should be acknowledged that output per worker raises a major econometric concern regarding a potential endogeneity problem.[4] Amiti and Wei (2009) suggested an alternative way to address the potential endogeneity problem — using value added per worker measured by taking the difference between output and intermediate materials and services, divided by employment. This measure of value added per worker, however, does not qualitatively affect the main results presented in the next section.

4.5 Empirical Evidence from Thailand's Establishment-Level Data

This section presents my empirical evidence regarding a relationship between outsourcing and labour productivity based on Thailand's establishment-level data. My presentation strategy follows that portrayed in Chapter 3 — the fitted plots, for the purpose of presentation simplicity, are constructed by aggregating the establishment-level data at the four-digit ISIC levels whereas the estimates are obtained by employing the full establishment-level dataset comprising more than 8,000 establishments. It should be further highlighted that the proxy of labour productivity (y_{li}) is represented in the logarithmic form.

Consistent with a large number of studies that asserted labour productivity gains from outsourcing activities,[5] my empirical evidence reveals that workers deployed in Thailand's manufacturing industries tend to benefit from fragmentation of production stages and outsourcing. As portrayed in Figure 4.1, labour productivity is positively correlated with outsourcing intensity with the estimated

FIGURE 4.1
A Fitted Plot of Outsourcing Intensity and Labour Productivity

$$y_{li} = 8.6211 + 3.7148^{***} O_i^T + u_i$$
$$(.1185) \quad (.1379)$$
$$N = 8811 \; ; \; F = 725.9^{***} \; ; \; R^2 = .0761$$

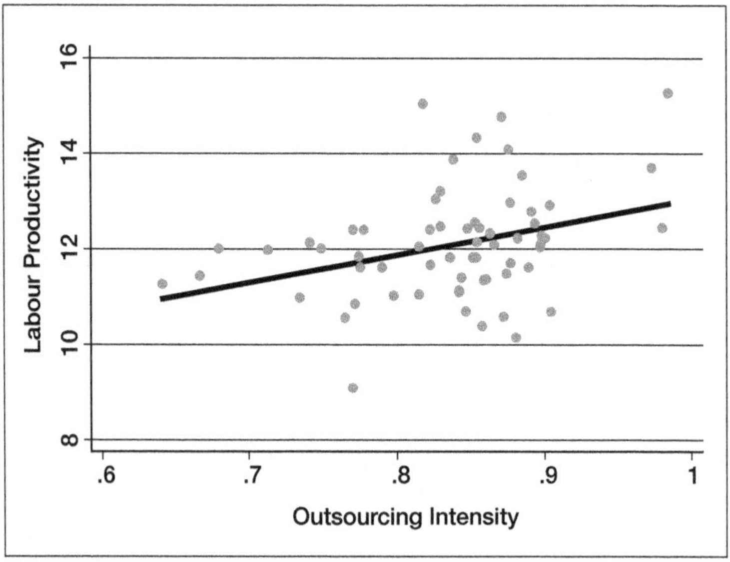

Notes: (1) *** represents statistical significance at 1 per cent; and (2) standard errors reported in parentheses.

coefficient equal to 3.72 and statistically significant at the 1 per cent level.

How does outsourcing cause a boost in labour productivity? A rise in labour productivity may be attributable to either compositional or structural changes (Amiti and Wei 2009). Outsourcing enables a firm to trim relatively inefficient stages of production by procuring intermediate inputs from outside providers who can undertake these production activities more productively. As a result, a firm becomes more concentrated on the production stages in which it has comparative advantage, and thus these compositional changes enhance productivity of remaining workers. For instance, Thai manufacturers have increasingly contracted out human resources and information technology (IT) services departments to outside contractors. These compositional changes in in-house production imply that the existing workers they employ will be relocated to and hence more specialized in their core-competent activities, ultimately enjoying higher productivity.

In addition, outsourcing prompts structural changes in such a way that it provides better access to a pool of intermediate inputs. Cheaper prices, greater variety and higher quality of intermediate materials and services that are more available to a firm trigger a shift in labour productivity. For example, procuring computing and IT services tends to augment labour efficiency in that it motivates the existing labour to learn a new software package and how to improve production activities.

The empirical results elaborated thus far hint at labour productivity gains emanated from production fragmentation or outsourcing. The next step is to investigate the establishment-specific characteristics which may have a role to play in the mechanism through which outsourcing triggers a shift in labour performance. Based on the review of literature enumerated in Section 4.2, there are two establishment-specific characteristics — outsourcing locations and foreign ownership — of interests.

The existing empirical research seems to suggest that two notions of outsourcing — one loosely defined as the use of intermediate inputs supplied by either domestic or foreign providers and the other confined mainly to imported intermediate inputs — contributes to the observed labour productivity growth. A natural question arises: Which types of outsourcing — e.g. general or international outsourcing — have greater effects on labour productivity changes? Chongvilaivan and Hur (2011) exerted the first attempt to account for the role of outsourcing location in explaining a shift in labour productivity using the six-digit NAICS U.S. manufacturing industries data. They find that the labour productivity effects of general outsourcing are more pronounced than those of international outsourcing. Their results give rise to the role played by domestic outsourcing in explaining changes in labour productivity.

Figures 4.2 and 4.3 divide the dataset into two groups: one that engages in international outsourcing, and the other that procures intermediate materials and services merely from domestic providers. The results are strikingly consistent with those of Chongvilaivan and Hur (2011). As portrayed in Figures 4.2 and 4.3, labour productivity gains are observed in both types of establishments since both fitted plots exhibit upward sloping with positive, statistically significant coefficients. Furthermore, the

FIGURE 4.2
Fitted Plots of Outsourcing Intensity and Labour Productivity among International Outsourcing Establishments

$$y_{li} = 9.7085 + 4.2708^{***}O_i^T + u_i$$
$$(.2671) \quad (.2996)$$
$$N = 1418 \; ; \; F = 203.15^{***} \; ; \; R^2 = .1255$$

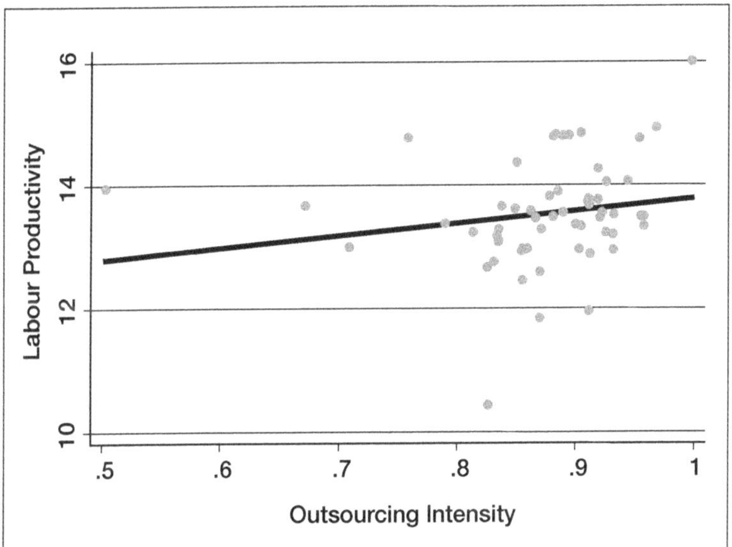

Notes: (1) *** represents statistical significance at 1 per cent; and (2) standard errors reported in parentheses.

positive effects on labour productivity are more pronounced for the establishments that contract out their production stages to both domestic and foreign partners (4.2708 > 2.9920). As did Chongvilaivan and Hur (2011), the empirical evidence presented in this section substantiates the roles of domestic outsourcing in explaining labour productivity growth.

FIGURE 4.3

Fitted Plots of Outsourcing Intensity and Labour Productivity among Domestic Outsourcing Establishments

$$y_{li} = 8.9845 + 2.9920^{***}O_i^T + u_i$$
$$(.1449) \quad (.1720)$$
$$N = 5146 \; ; \; F = 302.69^{***} \; ; \; R^2 = .0556$$

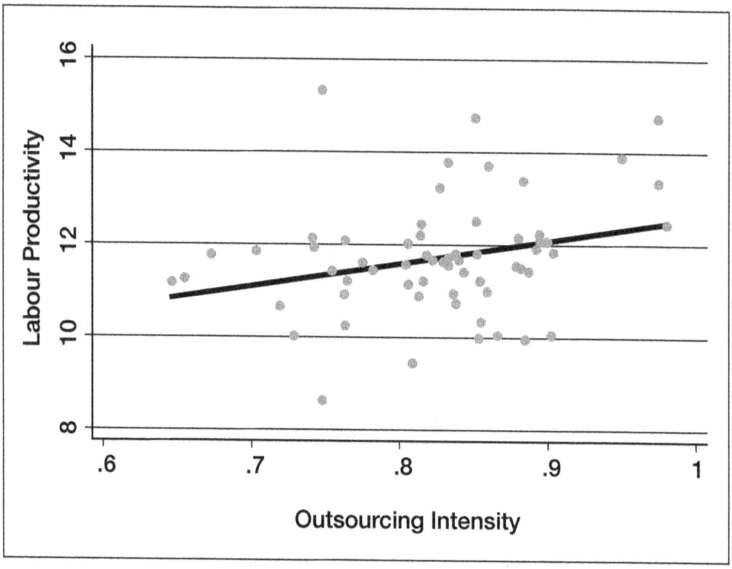

Notes: (1) *** represents statistical significance at 1 per cent; and (2) standard errors reported in parentheses.

Next I attempt to account for foreign ownership that may also result in different effects of outsourcing on labour productivity. Again, I divide the dataset into two groups: foreign-owned and locally-owned establishments.

As shown in Figures 4.4 and 4.5, a positive, statistically significant, correlation between outsourcing intensity and labour productivity can be observed in both foreign-

FIGURE 4.4
Fitted Plots of Outsourcing Intensity and Labour Productivity among Foreign-Owned Establishments

$$y_{li} = 10.9323 + 3.2361^{***}O_i^T + u_i$$
$$(.3694) \quad (.4141)$$
$$N = 808 \; ; \; F = 61.06^{***} \; ; \; R^2 = .0704$$

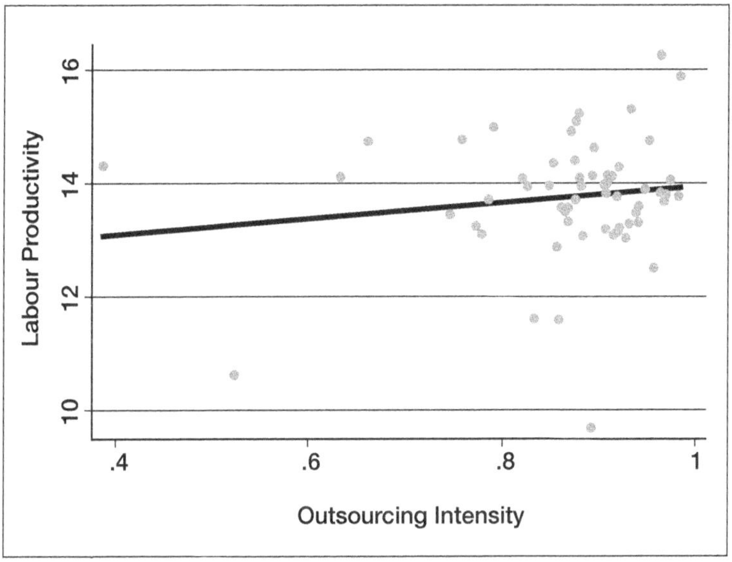

Notes: (1) *** represents statistical significance at 1 per cent; and (2) standard errors reported in parentheses.

and locally-owned firms. Moreover, the effects on labour productivity seem to be more pronounced, though negligibly, for locally-owned establishments (3.2361 < 3.3080).

My empirical results reveal that foreign ownership may have merely a limited role to play in explaining a linkage between outsourcing activities and labour productivity improvements in Thailand's manufacturing industries and

FIGURE 4.5

Fitted Plots of Outsourcing Intensity and Labour Productivity among Locally-Owned Establishments

$$y_{li} = 8.8771 + 3.3080^{***}O_i^T + u_i$$
$$(.1376) \quad (.1624)$$
$$N = 5756 \; ; \; F = 414.96^{***} \; ; \; R^2 = .0673$$

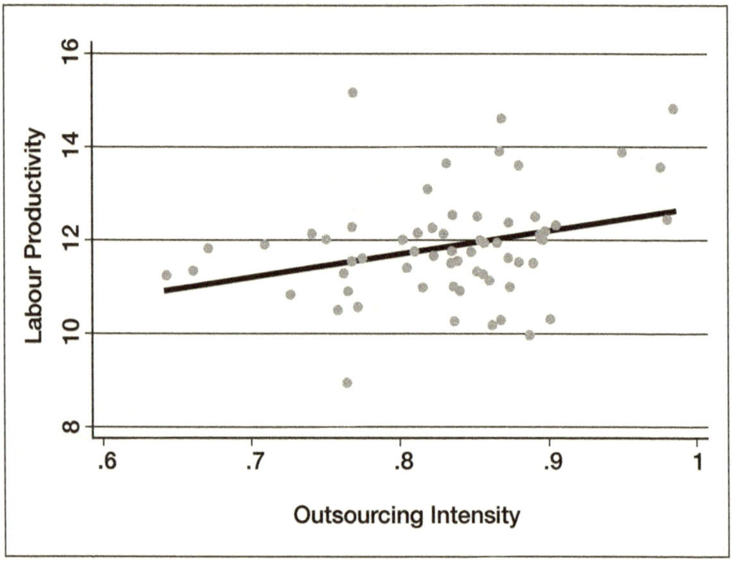

Notes: (1) *** represents statistical significance at 1 per cent; and (2) standard errors reported in parentheses.

therefore are in contrast with those of Girma and Görg (2004) who used UK establishment-level data and found that labour employed in a foreign-owned establishment tends to reap greater benefits in terms of an increase in its efficiency from outsourcing activities than those hired by locally-owned one. These contrasting results may be attributed to the fact that an international production network to which a foreign-owned establishment possesses access pertains to skill-intensive, high-tech production, but the production technology adopted by Thai manufacturers is relatively backward. In this sense, the existing labour would take considerable time to learn new production techniques and technology (e.g. new software, materials, management and designs), and hence its productivity improvements are unlikely to prevail at least in the short run.

4.6 Concluding Remarks

The present chapter examines the effects of production fragmentation on labour productivity and hence sheds further light on more intuitive understandings of its effects on the domestic economy. A brief review of the existing studies suggests that outsourcing in general improves labour performance and productivity; its merit, however, requires cautious contemplation since labour productivity improvements outsourcing entails hint crucially upon a number of factors, including outsourcing typology, skill levels of labour and locations of outsourcing and ownership, among many others.

My primal approach to productivity analyses based on a CES production function conceptually shows that there are three channels through which outsourcing alters labour productivity: *factor-productivity effect*, *technology effect* and *value-added effect*. The first channel captures the factor-augmenting technological effect of outsourcing — the effect that directly enhances productivity of a particular production factor. The second accounts for its effect that prompts a shift in overall productivity of a firm.

The last channel refers to productivity improvements in terms of higher value added a firm materializes. One can also discern that the last two channels affect productivity of all production factors uniformly.

This chapter then presents empirical evidence using Thailand's establishment-level data. The estimates reveal that labour employed in Thailand's manufacturing industries enjoys productivity gains from production fragmentation and outsourcing, and are therefore consistent with the existing findings. Two explanations may rationalize the observed boosts in labour efficiency. First, outsourcing enables the existing workers employed in-house to reap benefits from compositional changes — labour productivity improvements as the production structures become more concentrated on core-competent activities. In addition, structural changes associated with outsourcing activities may also trigger a surge in labour productivity levels since a firm that pertains to outsourcing, in principle, has better access to a cheaper pool, more varieties, and higher quality of intermediate materials and services than does the other which internalizes production activities.

A breakdown of samples into ones which procure intermediate inputs from both domestic and foreign sources and the others which rely exclusively on domestic providers further reveals that outsourcing locations matter. Labour productivity gains are significantly more pronounced for an establishment which pertains to international outsourcing. This finding, accordingly, confirms those of Chongvilaivan and Hur (2011). I then extend analyses by dividing the dataset into foreign- and locally-owned establishments. This extension produces the result contrasting with those of Girma and Görg (2004) that the effects of outsourcing on labour employed in locally-owned establishments are slightly more intensified than those in foreign-owned ones. This may be explained by the fact that the existing labour takes substantial time to learn skill-intensive, high-tech intermediate inputs (e.g. new software, advanced components and better product designs) the international production network offers, and hence more significant labour productivity gains are, at least in the short run, unlikely to be observed in foreign-owned establishments.

Notes

1. The effects of outsourcing on Thailand's labour markets have been pushed forward by Chongvilaivan and Thangavelu (2008). However, their focus is on whether and how outsourcing widens wage differentials between skilled and unskilled workers using the framework of a translog cost function whereas the present chapter aims to investigate the impacts on overall productivity of labour using a relatively straightforward empirical methodology.
2. See Jones and Kierzkowski (2001) for more detailed discussions of a theoretical treatment of production fragmentation and outsourcing.
3. See Section 4.4 for more detailed discussions regarding labour productivity measurements.
4. See Chongvilaivan and Hur (2011) for detailed discussions regarding the treatment of the potential endogeneity problem.
5. See Section 4.2 for a review of the existing studies.

5
Skills Upgrading: Winners or Losers?

5.1 Introduction

A growing number of recent studies dealing with the economic impacts of production fragmentation on the labour market have increasingly weighed in on the phenomenon of a widening income gap between skilled and unskilled workers in developed countries.[1] In these studies, the evidence pointed to the international dimension of production fragmentation — particularly, international outsourcing or the uses of intermediate imports — as a key catalyst of this trend in that technical advances in information technology and closer trade ties to the global market have entailed surges in outsourcing of less skill-intensive activities to unskilled labour-abundant countries, thereby spurring the worsening wage inequality in terms of exporting unskilled jobs to developing countries (Amiti and Wei 2009). A theoretical relationship between imports of intermediate inputs and a shift in relative demand for skilled workers observed in the industrialized economies lies with the standard Heckscher-Ohlin (H-O) Theorem. In this well-known theoretical framework, the industrialized economies where skilled labour is typically well-endowed are predicted to specialize more skill-intensive production activities and make use of less skill-intensive intermediate imports from developing countries where unskilled labour is relatively abundant. Given the perfectly competitive labour market, the reallocation of resources toward skill-intensive sectors will ultimately shift the relative demand for skilled workers in industrialized economies.

Recent developments, nonetheless, indicated that developing countries are also affected by fragmentation of the production. In fact, outsourcing activities have increased substantially among the Southeast Asian countries (Ahn et al. 2008). Even though Japan has long been an important source of intermediate inputs for these regions, the evidence shows that the Southeast Asian countries are also emerging as crucial source countries for intermediate inputs. Concurrently, it has been observed that the Southeast Asian countries have experienced staggering increases in intermediate imports from China and other key Southeast Asian members, especially Indonesia, Malaysia, the Philippines and Thailand. These observations therefore highlight that firms operating in this region procure intermediate inputs from both local and foreign suppliers.

In this regard, the idiosyncratic nature of outsourcing activities in this region forces me to approach this empirical question in a non-standard way. The impacts of production fragmentation on labour markets in developing Southeast Asia must pertain not only to the imports of intermediate inputs, but also to contracting out production activities at arm's length locally. The notion of outsourcing employed in this paper, accordingly, takes the general outsourcing definition as in Chongvilaivan et al. (2009) and thus abstracts from the outsourcing definition conventionally employed in the literature.

The mechanism through which outsourcing alters the relative demand for skilled workers in this chapter departs from what the international trade literature suggests. It rests with the labour economics arguments as pushed forward by Chongvilaivan and Hur (2011). In this chapter, I extend the simple CES production framework developed in the earlier chapters and show that widening wage differentials between skilled and unskilled workers take root in the skill-biased effects of outsourcing. That is, labour productivity improvements from outsourcing activities that are in favour of skilled workers trigger an outward shift in relative demand for skilled workers.

Then, I empirically examine the impacts of outsourcing activities on relative demand for skilled workers using Thailand's establishment-level manufacturing industry data. As conventionally employed in the existing studies like Feenstra and Hanson (1996*a*, 1996*b*, 1999), Head and Ries (2002), Hsieh and Woo (2005), and Hijzen et al. (2005), the relative demand for skilled workers is measured by the share of the skilled worker payroll. It should also be emphasized that skilled labour is proxied by non-production workers such as managers, directors, laboratory and research staff, and administrative supervisors, as suggested by Gujarati and Dars (1972).

I find that outsourcing of intermediate inputs has no significant effects on relative demand for skilled workers in Thailand's manufacturing sector. There are at least three explanations of the weak evidence. Firstly, since the index of outsourcing intensity includes not only the intermediate imports, but also materials and services intermediates supplied by local partners, it is possible that domestic and international outsourcing have opposing effects on the relative skilled labour demand. Nevertheless, a breakdown of samples according to the locations of outsourcing reveals that the effects on relative skilled labour demand remains insignificant in all sub-samples. The second possibility that rationalizes the weak evidence is that outsourcing uniformly affects productivity of skilled and unskilled workers, leaving the relative demand for skilled workers unaffected. Chongvilaivan et al. (2009) pointed out the third possibility that outsourcing typology matters. Even though the second explanation seems plausible from the theoretical point of view, it deserves further empirical investigations since each outsourcing type may not identically affect all skill groups. Rather, it may bring about contrasting effects on relative skilled labour demand. For instance, if different types of outsourcing render different impacts on the relative demand for skilled workers, the net effect will be cancelled out. I will relegate further examination of outsourcing typology to the next chapter.

A breakdown of the establishment samples according to their foreign ownership sheds further light on a relationship between outsourcing activities and widening wage inequality between skilled and unskilled workers in Thailand's manufacturing sector. The empirical evidence indicates that foreign establishments tend to specialize in less skill-intensive production activities and contract out more skill-intensive production at arm's length, thereby prompting a plunge in the shares of the skilled worker payroll. These results imply that Multinational Enterprises

(MNEs) operating in Thailand's manufacturing sector reap benefits from unskilled labour cost savings by concentrating on less skill-intensive production activities.

The rest of this chapter is organized as follows. Section 5.2 provides a brief survey of the existing empirical research on a relationship between outsourcing and widening wage inequality. Section 5.3 develops a simple CES production function framework and derives a condition under which outsourcing activities drive an outward shift of relative demand for skilled workers. Section 5.4 discusses the measurement of relative skilled labour demand. Section 5.5 presents and analyses the empirical results based on Thailand's establishment-level data. Section 5.6 concludes.

5.2 Review of the Literature

The past two decades witnessed a remarkable increase in production fragmentation together with a sharp plunge in the wages of unskilled workers, both in real terms and relative to the wages of skilled workers, especially in advanced countries like the United States (U.S.) and the European Union (EU). These observations give rise to considerable debate over whether and how production fragmentation, on top of the existing argument of skill-biased technological progress, has contributed to widening wage inequality between skill groups in the labour markets.

A number of research studies put emphasis on the roles of globalization in shifting the relative demand for skilled workers and are thus concerned with the cross-border element of production fragmentation, so called international outsourcing. Feenstra and Hanson (1996*a*) are the first to push forward the theoretical propositions that account for these empirical observations, building upon the standard Heckscher-Ohlin (H-O) Theorem. In their model, the underlying assumption that drives their main results hints on international differences in relative wages of skilled and unskilled workers in two regions — namely, more skill-intensive North and less skill-intensive South. They showed that international openness enables the North to specialize more skill-intensive production stages through the uses of less skill-intensive intermediate imports from the South and ultimately prompts a shift in the relative demand for skilled workers in the North. The uses of outside providers ultimately trigger shifts in both intra-firm and intra-industry relative demand for skilled workers, thereby rationalizing the rising wage inequality phenomenon.

Among the very first studies that empirically examined the extent to which imports of intermediate inputs impinge on the relative demand for skilled workers is that by Feenstra and Hanson (1996*b*). They showed that international outsourcing can account for 15–33 per cent of the relative increases in relative wages of skilled workers using U.S. manufacturing industry data from 1979 to 1990. In their later study, Feenstra and Hanson (1999) examined two competing hypotheses of the widening wage gap in the U.S. labour market — international outsourcing and skill-biased technological progress — and revealed that technological changes that augment relatively skilled workers also have a pivotal role to play in explaining the observations.

Following the publications of these two seminal papers, several authors have replicated their results employing data from various developed countries. Anderton and Brenton (1999), for example, estimated the impacts of international outsourcing on the wage-bill shares and relative employment of unskilled workers in the United Kingdom (UK) manufacturing sector. Their empirical estimates revealed that outsourcing to low-wage countries contributes negatively to relative demand for unskilled workers (and hence positively to relative demand for skilled workers). Hijzen et al. (2005) departed from the existing studies by developing an index of international outsourcing using import-use matrices of input-output (IO) tables of the UK manufacturing industries for the period of 1982–96. Their results are consistent with those of Anderton and Brenton (1999), confirming that international outsourcing has strong, negative effects on the relative demand for unskilled workers. Following the empirical approach pioneered by Feenstra and Hanson (1999), Hijzen (2007) further showed that international outsourcing contributed significantly to wage inequality even though skill-biased technological changes are also a predominant driver of this phenomenon in the UK industries during 1993–98.

Several recent studies also drew similar conclusions that international outsourcing has contrasting impacts on wages across skill groups in various industrialized economies using micro-level data. The merits of exploiting micro-level data rest with avoiding estimation problems that are common drawbacks of industry-level studies, such as aggregation biases, potential endogeneity biases and deprived skill definitions. In particular, Geishecker and Görg (2008) empirically investigated the link between international outsourcing and wages across different skill groups.[2] Based on a large household panel combined with industry-level information on industries' outsourcing activities in Germany for the years 1991 to 2000, their evidence indicated that international outsourcing has a remarkable impact on wage differentials. Specifically, their estimates predicted that a 1 percentage point increase in outsourcing intensity trims the wages for the lowest-skilled workers by up to 1.5 per cent while raising those for high-skilled workers by up to 2.6 per cent.

Head and Ries (2002) likewise examined the influence of Japanese multinationals' employment in low-income countries on domestic skill upgrading using a panel of 1,070 Japanese firms. They showed that those affiliates' employment in low-income countries makes a firm more dependent on relatively less skill-intensive intermediate imports, replaces domestic employment of unskilled workers, and therefore contributes positively to domestic skill upgrading.

Furthermore, Hsieh and Woo (2005) centred on the impacts of outsourcing to China, on the widening wage gap in Hong Kong using the micro-level data for the period 1971–96. They found evidence of strong and persistent shifts in the relative demand for skilled workers in Hong Kong since the 1980s during which outsourcing to China became prominent. They further showed that 15 per cent of the observed demand shifts is attributed to reallocation of resources towards services related to outsourcing while roughly another 15 per cent goes to the reallocation of unskilled workers within manufacturing industries.

Even though an overwhelming amount of research unanimously pointed to the empirical evidence of a positive relationship between international outsourcing and relative demand for skilled workers in advanced economies, the evidence in developing countries is rather mixed and less clear-cut. For instance, Fajnzylber and Fernandes (2009) empirically compared the impacts of imported inputs on the skilled labour demand employing two firm-level datasets in two developing countries, Brazil and China.[3] They found evidence that the use of imported inputs in the two countries has different effects on the skilled labour demand. While Brazilian firms that pertain to intermediate imports tend to have higher demand for skilled workers, the opposite relationship is observed among Chinese firms. A negative relationship between international outsourcing and skilled labour demand in China is perhaps attributable to the fact that Chinese firms have comparative advantages in less skill-intensive production activities and thus rely on the imports of more skill-intensive inputs.

Thangavelu and Chongvilaivan (2010) abstracted from the existing studies by pondering two types of international outsourcing, materials and services outsourcing, using Thailand's industry-level panel data for the period of 1999–2003, aggregated from establishment-level data at the four-digit industrial classification level. Their evidence revealed that materials outsourcing is negatively correlated with the relative demand for skilled workers whereas services outsourcing tends to shift the skilled labour demand. These results may suggest that in Thailand's manufacturing industries services activities are likely to be more skill-intensive while materials production tends to be less skill-intensive.

A body of recent research extends the notion of production fragmentation to include not only its international element, but also its domestic element — general outsourcing as in Abraham and Taylor (1996) — since it builds upon research inquiries into the labour productivity effects of production fragmentation. As discussed in Chapter 4, several past studies have found strong evidence of labour productivity gains from general outsourcing, such as and Girma and Görg (2004) and Görzig and Stephan (2002). As empirically shown by Chongvilaivan and Hur (2011), production fragmentation will ultimately result in a shift in the relative demand for skilled labour. That is, production fragmentation is skill-biased — skilled labour reaps more benefits from production fragmentation in terms of its productivity improvements through specialization, scale economies, distributing fixed costs and access to a larger pool of cheaper and improved intermediate inputs than does unskilled labour.

Few studies have pushed forward evidence of general outsourcing as a determinant of rising wage inequality. Chongvilaivan et al. (2009), for instance, employed the six-digit North American Industry Classification System (NAICS) U.S. manufacturing industries and contemplated a breakdown of outsourcing into three categories: upstream materials outsourcing, downstream materials outsourcing and services outsourcing. They found that downstream materials and services outsourcing contribute significantly to an upward shift in the relative skilled labour demand, whereas upstream materials outsourcing does not. Geishecker and Görg (2008) extended their own analyses of a relationship between international outsourcing and labour demand by adding the index of international outsourcing into their baseline empirical framework. They showed a contrasting result that domestic outsourcing has merely insignificant impacts on demand for labour at all skill levels.

5.3 Production Fragmentation and Wage Inequality

This section examines how outsourcing activities shift the relative demand for skilled workers, employing the simple CES production function framework as put forward by Egger and Egger (2006) and Chongvilaivan and Hur (2011). Recall that in Chapters 3 and 4 there are two production factors entering the production function, capital and labour. In this specification, labour is implicitly assumed to have an identical skill level. In order to develop further analyses of wage differentials, this assumption must be relaxed. Therefore, I assume that there are two skill groups of labour an establishment i employed, namely skilled workers (L_{Hi}) and unskilled workers (L_{Li}), and hence the CES production function as appeared in Equation (3.3) can be modified as follows:

$$V_i = e^{\alpha + \gamma O_i}\{(e^{\beta_K O_i}K_i)^{\rho} + (e^{\beta_H O_i}L_{Hi})^{\rho} + (e^{\beta_H O_i}L_{Li})^{\rho}\}^{r/\rho} \qquad (5.1)$$

In Equation (5.1), β_H and β_L capture skilled and unskilled labour-augmenting technological changes triggered by outsourcing activities. By differentiating Equation (5.1) with respect to L_{Hi} and L_{Li}, one can straightforwardly obtain the relative demand for skilled workers represented by the ratio of marginal productivity of skilled labour (V_{Hi}) to that of unskilled labour (V_{Li}).

$$\frac{V_{Hi}}{V_{Li}} = e^{\rho(\beta_H - \beta_L)O_i}(L_{Hi}/L_{Li})^{\rho-1} \tag{5.2}$$

Next, the elasticities of the relative skilled labour demand with respect to outsourcing activities (η_{wo}) can be derived by logarithmically differentiating Equation (5.2) with respect to the index of outsourcing intensity (O_i).

$$\eta_{wo} \equiv \frac{d\ln(V_{Hi}/V_{Li})}{d\ln O_i} = \rho(\beta_H - \beta_L)O_i \tag{5.3}$$

Equation (5.3) implies that the impacts of outsourcing on relative demand for skilled labour depend crucially on whether its labour-augmenting effect is skill-biased (e.g. whether $\beta_H > \beta_L$). If labour productivity gains from outsourcing activities are in favour of skilled labour ($\beta_H > \beta_L$), outsourcing activities are positively correlated with the wage gap between skilled and unskilled workers.

5.4 Measurement of Relative Skilled Labour Demand

In order to understand the essence of the change in the composition of manufacturing employment, it is indispensable to clearly identify the definitions of skilled and unskilled labour as well as the measurement of relative skilled labour demand. There are a number of approaches to skill classifications, such as education levels, training, skills, experience, and the nature of activities engaged. Much research such as Feenstra and Hanson (1996*a*, 1996*b*, 1999), Anderton and Brenton (1999), Head and Ries (2002), Hsieh and Woo (2005), Hijzen et al. (2005), and Chongvilaivan et al. (2009), among many others, commonly defined unskilled and skilled workers according to two types of an establishment's activities — production and non-production activities — they are engaged in. Production workers are those deployed in production activities such as packaging, assembling, inspecting, repair, maintenance and watchman services, while non-production workers are those engaged in non-production activities such as factory supervision, office functions as well as financial, legal, professional and technical services. Since the production activities in general are less skill-intensive than the non-production activities, it is conventionally accepted that production workers are unskilled and non-production workers are skilled (Gujarati and Dars 1972).

This approach to labour skill classifications has the best fit with my dataset. In the *2003 Manufacturing Industry Survey*, the National Statistical Office (NSO) classified workers engaged in a manufacturing establishment into two categories: production and non-production workers. The former refers to those who were employed in the production or other related activities such as workers in factories, caretakers, machine tenders, machine controllers and assemblers while the latter includes administrative, technical and clerical personnel such as managers, directors, laboratory and research workers, and administrative supervisors and the like. Throughout this chapter, therefore

I measure skilled and unskilled workers as non-production and production workers, respectively.

Having defined the measurement of skilled and unskilled workers, I next turn to the discussions of the relative skilled labour demand measurements. The share of skilled workers in total employment is perhaps the simplest proxy and has been widely employed by the previous studies such as Machin et al. (1996). The other alternative measure — the ratio of skilled to unskilled wages, representing the relative wage differentials — was also employed by recent studies such as Egger and Egger (2006) and Chongvilaivan and Hur (2011). These measures of relative skilled labour demand, however, are less satisfactory from the theoretical point of view since it is not derived from a firm's profit maximization condition as economic theories suggest (Anderton and Brenton 1999).

A conventional measure of the relative skilled labour demand is obtained from the short-run cost function, where the levels of capital and output are given. As in Chongvilaivan et al. (2009), I without loss of generality assume that the cost function takes a logarithm form as follows:[4]

$$C_i(w_{Hi}, w_{Li}) = \min_{L_{Hi}, L_{Li}} w_{Hi}L_{Hi} + w_{Li}L_{Li}$$
$$\text{subject to } Y_i = F_i(L_{Hi}, L_{Li}, K_i, X_i) \qquad (5.4)$$

where the subscript i denotes an establishment i; c_i is total wage bill; L_{Hi} is the number of skilled workers; w_{Hi} is the skilled labour wage rate; L_{Li} is the number of unskilled workers; w_{Li} is the unskilled wage rate; Y_i is the level of output produced; K_i is quasi-fixed capital; and X_i is the external shift factors such as outsourcing indices and technological changes. By using the Envelope Theorem, one can easily show that the wage share of the skilled worker payroll (WS_{Hi}) can be derived by logarithmically differentiating Equation (5.1) with respect to w_{Hi}. That is,

$$WS_{Hi} = \frac{\partial \ln c_i(w_{Hi}, w_{Li})}{\partial \ln w_{Hi}} \frac{w_{Hi} L_{Hi}}{c_i(w_{Hi}, w_{Li})} \tag{5.5}$$

It is conventionally known that the share of the skilled worker payroll (WS_{Hi}) is an expression of the relative skilled labour demand. It captures not only the relative employment, but also relative wages. This measure has by and large been employed to empirically investigate the impacts of outsourcing on wage inequality between skilled and unskilled workers by various studies such as Hanson and Harrison (1999), Anderton and Brenton (1999), Dell'mour et al. (2000), Geishecker (2002), Hsieh and Woo (2005) and Chongvilaivan et al. (2009).

5.5 Empirical Evidence from Thailand's Establishment-Level Data

This section empirically examines a relationship between production fragmentation and the relative demand for skilled labour using Thailand's establishment-level manufacturing industry data. As done in the previous chapters for the sake of presentation simplicity, the fitted plots are based on the establishment-level data aggregated at the four-digit ISIC level, yielding sixty-three manufacturing industries, while the parameter estimates are obtained from the full

FIGURE 5.1

A Fitted Plot of the Outsourcing Intensity and the Shares of the Skilled Worker Payroll

$$WS_{Hi} = .2702^{***} + .0114O_i^T + u_i$$
$$(.0126) \qquad (.0145)$$
$$N = 3998 \; ; \; F = .62 \; ; \; R^2 = .0002$$

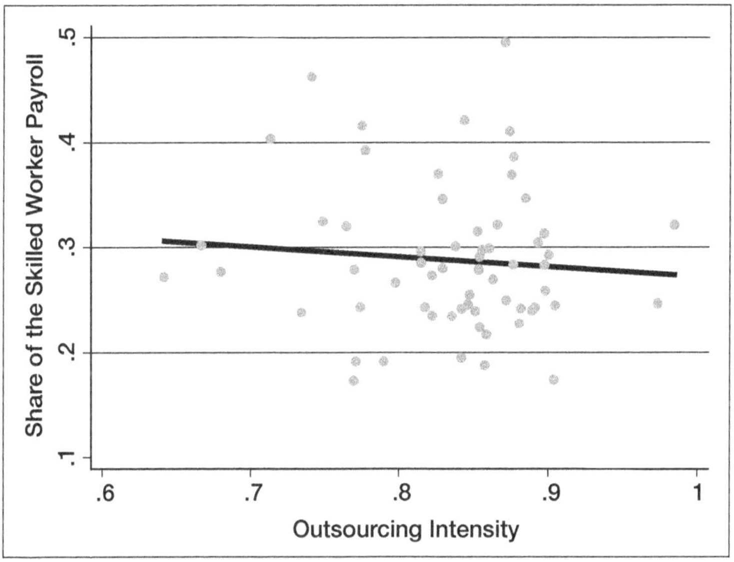

Notes: (1) *** represents statistical significance at 1 per cent; and (2) standard errors reported in parentheses.

establishment-level samples. In the dataset, there are 3,998 establishments whose information on skilled worker payrolls and outsourcing intensity is available.

As portrayed in Figure 5.1, the fitted plot between outsourcing intensity and the shares of the skilled worker payroll is nearly flat and the coefficient of O_i^T, though positive, is statistically insignificant. This implies that outsourcing activities in Thailand's manufacturing industries in general have no significant effects on a shift in relative skilled labour demand. There are at least three explanations of the weak evidence observed in our dataset. Firstly, the absence of evidence may be attributed to contrasting effects of domestic and international outsourcing on relative demand for skilled workers as empirically shown by Geishecker and Görg (2008).

Figures 5.2 and 5.3 further explore this argument of weak evidence of a relationship between outsourcing activities and relative skilled labour demand in Thailand's manufacturing industries by dividing the sample establishments into two categories: international outsourcing establishments and domestic outsourcing establishments. International outsourcing establishments refer to those which employ intermediate imports while domestic outsourcing establishments pertain merely to intermediate input procurement from local sources.

In Thailand's manufacturing industries, contracting out production activities at arm's length pertains to two contrasting effects on relative demand for skilled workers. On the one hand, productivity gains may be in favour of skilled labour since several studies pointed to skill-biased productivity effects of outsourcing as a catalyst of an outward shift in the relative skilled labour demand, including Egger and Egger (2006) and Chongvilaivan and Hur (2011). The standard Heckscher-Ohlin (H-O) Theorem, on the other hand, suggests that outsourcing could also be in favour of unskilled labour in developing countries where unskilled workers are relatively well-endowed. Firms tend to specialize unskilled labour-intensive production activities and employ imports of skilled labour-intensive intermediates from developed countries, thereby augmenting relative demand for unskilled labour employed in-house. Given these opposing effects, it seems plausible that the insignificant linkages between outsourcing activities and relative skilled labour demand are observed in Thailand's manufacturing industries.

FIGURE 5.2

Fitted Plots of Outsourcing Intensity and the Shares of the Skilled Worker Payroll among International Outsourcing Establishments

$$WS_{Hi} = \underset{(.0285)}{.2836^{***}} + \underset{(.0316)}{.0093} O_i^T + u_i$$

$$N = 1388 \; ; \; F = .09 \; ; \; R^2 = .0001$$

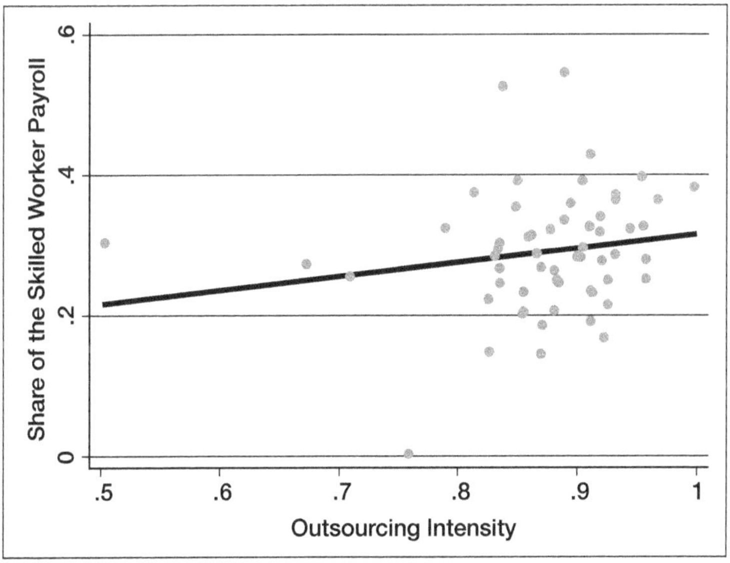

Notes: (1) *** represents statistical significance at 1 per cent; and (2) standard errors reported in parentheses.

FIGURE 5.3

Fitted Plots of Outsourcing Intensity and the Shares of the Skilled Worker Payroll among Domestic Outsourcing Establishments

$$WS_{Hi} = \underset{(.0141)}{.2699^{***}} + \underset{(.0166)}{.0042}O_i^T + u_i$$

$$N = 2610 \; ; \; F = .06 \; ; \; R^2 = .0001$$

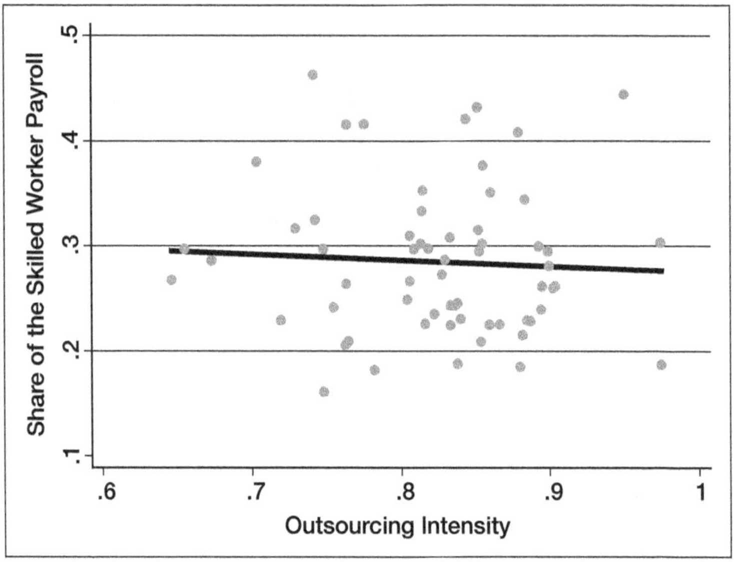

Notes: (1) *** represents statistical significance at 1 per cent; and (2) standard errors reported in parentheses.

The evidence, however, seems contradictory to the above presumption since the evidence exhibits no systematic differences in the impacts of outsourcing on the shares of the skilled labour payroll between two types of establishments. That is, outsourcing activities do not render significant impacts on the relative demand for skilled workers in both international and domestic outsourcing establishments since the coefficients of o_i^T are positive and statistically in both estimations.

Secondly, based on my CES production function framework, the weak evidence implies the absence of skill-biased labour productivity effects ($\beta_H = \beta_L$). That is, outsourcing activities uniformly affect skilled and unskilled labour productivity improvements and thus leave the relative demand for skilled workers unchanged. Last but not least, an alternative conjecture of the weak evidence perhaps rests with the roles of outsourcing typology as pioneered by Chongvilaivan et al. (2009). They found that outsourcing of materials and services exhibited contrasting effects on relative demand for skilled workers in the U.S. manufacturing industries. It is thus possible that outsourcing does not have identical productivity effects on each skill group. Rather, the weak evidence is caused by different types of outsourcing yielding contrasting productivity effects on skilled and unskilled workers. I will relegate thorough explorations of outsourcing typology roles in explaining widening wage inequality to next chapter.

Figures 5.4 and 5.5 view the mechanism through which outsourcing activities affect relative demand for skilled workers from the other angle. It breaks down the sample establishments into two categories: foreign and locally-owned establishments and thus provides further insights into foreign ownership as a factor that accounts for the effects of outsourcing on the wage differentials.

The interesting results are as follows. The coefficient of o_i^T is negative and statistically significant in the sub-samples with foreign-owned establishments, but it remains

FIGURE 5.4
Fitted Plots of Outsourcing Intensity and the Shares of the Skilled Worker Payroll among Foreign-Owned Establishments

$$WS_{Hi} = .3982^{***} - .1105^{**}O_i^T + u_i$$
$$(.0387) \qquad (.0432)$$
$$N = 825 \; ; \; F = 6.56^{**} \; ; \; R^2 = .0079$$

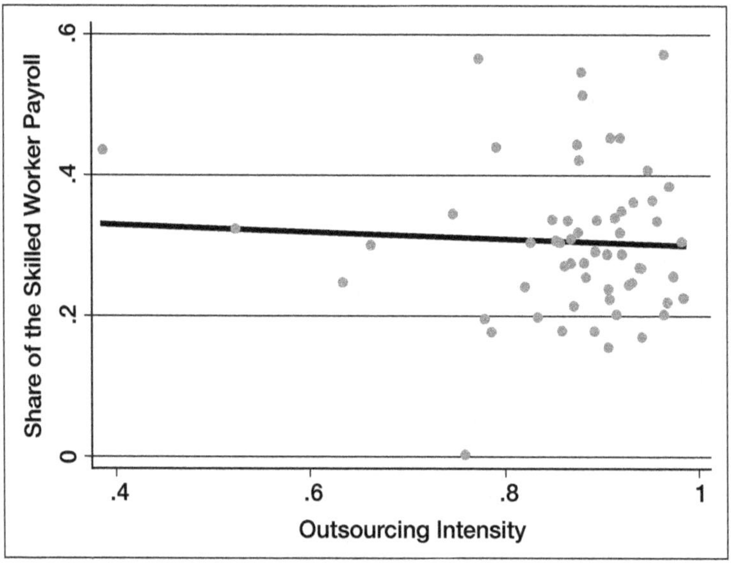

Notes: (1) *** represents statistical significance at 1 per cent; (2) ** represents statistical significance at 5 per cent; and (3) standard errors reported in parentheses.

statistically insignificant in the other sub-samples with locally-owned establishments. These results imply that foreign-owned establishments tend to contract out more skill-intensive intermediate inputs at arm's length and concentrate on less skill-intensive production activities in-house. This empirical evidence seems plausible given the

FIGURE 5.5

Fitted Plots of Outsourcing Intensity and the Shares of the Skilled Worker Payroll among Locally-Owned Establishments

$$WS_{Hi} = .2559^{***} + .0222O_i^T + u_i$$
$$(.0132) \qquad (.0154)$$
$$N = 3173 \; ; \; F = 2.09 \; ; \; R^2 = .0007$$

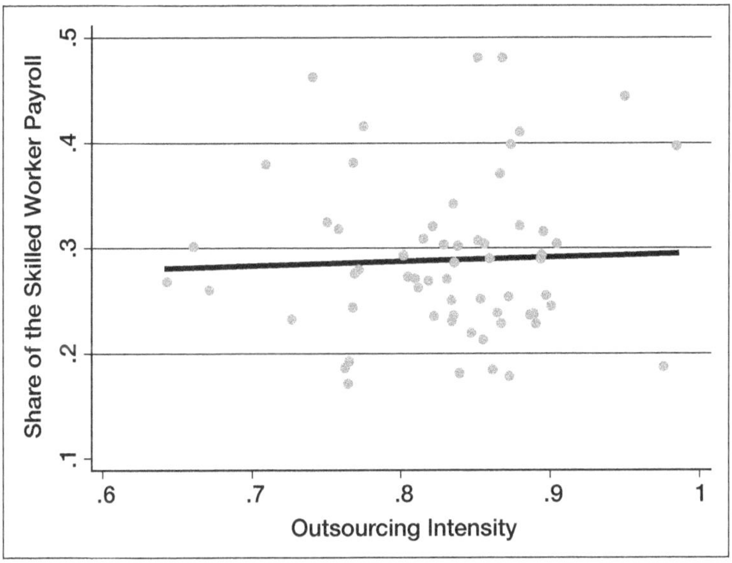

Notes: (1) *** represents statistical significance at 1 per cent; and (2) standard errors reported in parentheses.

fact that Thailand is a developing country and is thus well-endowed with unskilled workers. The MNEs that establish their affiliates in Thailand are able to exploit labour cost savings by undertaking less skill-intensive production activities in-house together with employing imports of more skill-intensive intermediate inputs. Outsourcing activities among foreign-owned establishments, therefore, put a downward pressure on relative demand for skilled workers in Thailand's manufacturing sector.

In a nutshell, empirical exercises presented in this section demonstrated weak evidence of a relationship between outsourcing activities and relative skilled labour demand in Thailand's manufacturing industries. A robustness check is carried out in two ways and further reveals the followings. First, a breakdown of the establishment samples according to locations of outsourcing activities substantiates the baseline results that outsourcing activities have merely a negligible role to play in shifting the relative skilled labour demand. In addition, the negative, statistically significant linkages between outsourcing activities and relative skilled labour demand are observed merely among foreign-owned establishments. This implies that MNEs operating in Thailand's manufacturing sector resort to outsourcing of the relatively skill-intensive intermediate inputs and carry out production of the less skill-intensive ones in-house.

5.6 Concluding Remarks

This chapter examines the effects of production fragmentation on relative demand for skilled workers. Using the simple CES production function framework in which the outsourcing intensity index is a shifting parameter, I show that outsourcing activities contribute to rising wage inequality when its labour productivity effects are skill-biased — that is, productivity improvements of skilled workers are more pronounced than those of unskilled workers. The empirical estimates, nevertheless, portray only weak evidence that outsourcing activities prompt a shift in relative skilled labour demand in Thailand's manufacturing industries. The evidence remains insignificant with a breakdown of the samples into two categories: (1) Establishments which employ both domestic and international outsourcing; and (2) Establishments which pertain merely to domestic outsourcing. At first glance, these results are surprising and seem to suggest that production fragmentation uniformly affects labour across skill groups and thus has a trifling role to play in contributing to a wage gap between skilled and unskilled workers. However, it should be emphasized that the index of general outsourcing employed in this chapter includes procurement of both intermediate materials and services. The aggregation of outsourcing activities points to the findings by Chongvilaivan et al. (2009), which highlight that a closer look into outsourcing typology is needed as different types of outsourcing activities may bring about the different impacts on the relative demand for skilled workers. For instance, if outsourcing of materials and services exhibits opposite effects on the shares of the skilled labour payroll, it is possible that the aggregate index of outsourcing exerts the nil effects. I will relegate further empirical investigations on outsourcing typology to the next chapter.

A breakdown of the samples into foreign- and locally-owned establishments sheds further light on intuitive insights into the impacts of outsourcing on the relative demand for skilled workers. I find evidence that the relative demand for skilled workers is negatively and significantly correlated with outsourcing activities only among the foreign affiliates even though the relationship remains weak among the locally-owned establishments. This may imply that MNEs operating in Thailand's manufacturing sector tend to contract out more skill-intensive intermediate inputs and internalize less skill-intensive production activities, ultimately entailing a downward shift in the relative demand for skilled workers within establishments and industries.

Notes

1. Following Feenstra and Hanson (1996*a*, 1996*b*, and 1999), several later studies examined the impacts of outsourcing on labour markets utilizing data retrieved from various industrialized economies such as Anderton and Brenton (1999) for the United Kingdom (UK), Geishecker (2002) for Germany and Hsieh and Woo (2005) for Hong Kong, among many others.
2. In contrast to most existing studies which broadly classified labour skill levels into non-production and production workers, Geishecker and Görg (2008) divided labour wages according to their education levels into three types: high-skilled labour (with the second stage of tertiary education), medium-skilled labour (with upper secondary education, post-secondary, non-tertiary education, or the first stage of tertiary education), and low-skilled labour (with lower secondary education or the second stage of basic education).
3. Besides the use of imported inputs, Fajnzylber and Fernandes (2009) also addressed the other two international economic activities including exports and foreign direct investment.
4. It should be highlighted that Chongvilaivan et al. (2009) made use of the translog cost function to derive econometric specifications.

6
Outsourcing Typology: Does It Matter?

6.1 Introduction

Much research has been devoted to attempts to empirically investigate the extent to which disintegrating production affects an economy. These include the effects on firm performance, factor productivity and labour markets. In these studies, the notion of production fragmentation has to do with the aggregate uses of materials and services inputs. However, recent developments witnessed a growing amount of evidence showing that there are different economic effects pertinent to contracting out of different types of intermediate inputs. This implies that a more refined treatment of outsourcing activities is indispensable for obtaining more accurate empirical results.

The role of outsourcing typology is not new. Several studies in various economies have pushed forward different types of outsourcing activities that exert different economic effects. There are at least three main approaches prevalently employed in the literature to take in heterogeneous effects of outsourcing typology.

Firstly, the nature of outsourced inputs (tangibles or services) is perhaps the most commonly utilized approach to separation of outsourcing types. A number of recent studies showed evidence that outsourced materials and services matter to the underlying economic impacts in contrasting ways. Görg and Hanley (2005), for instance, empirically examined a relationship between outsourcing activities and firm-level productivity using the firm-level data for the electronics industry in Ireland. They found that materials outsourcing contributes to total factor productivity growth whereas there is no such positive effect observed for services outsourcing. Using the same Irish dataset, Görg et al. (2008) attempted to control for firm-level heterogeneity and further revealed that potential productivity improvements from outsourcing of services inputs accrue only to exporters. Likewise, Amiti and Wei (2009) investigated productivity effects of international outsourcing of materials and services inputs in the United States (U.S.) manufacturing industries. They found rather contrasting evidence that outsourcing of both materials and services inputs rendered positive effects on productivity and accounted for around 10 and 5 per cent of observed productivity growth, respectively, in the period of 1992–2000. Their robustness checks further indicated that productivity gains from services outsourcing are robust across all specifications, but those from materials outsourcing are not.

Secondly, outsourcing activities are categorized according to their stages of production involved. For example, Chongvilaivan et al. (2009) broke down materials outsourcing into two types, namely upstream and downstream materials outsourcing using the six-digit NAICS U.S. manufacturing industry data. The former is measured by the shares of the total production costs taken up by the costs of intermediate parts and materials employed in the upstream production stage while the latter is measured by the cost shares of downstream activities such as reprocessing, repackaging and blending, in total production costs. They hypothesized that upstream activities are typically more skill-intensive than the downstream counterpart, and thus different types of outsourcing should have different effects on the relative demand for skilled workers. They found supporting evidence that downstream materials outsourcing is skill-biased and thus contributes to an outward shift in the relative demand for skilled workers while upstream materials outsourcing is not.[1]

Last but not least, Anderton and Brenton (1996) partitioned the notion of outsourcing into two sub-categories according to the locations of outsourcing providers, namely intermediate imports from low-wage countries and those from advanced industrialized countries using industry-level data of the United Kingdom (UK) manufacturing sector. They demonstrated that intermediates imports from low-wage countries are concerned with a decline in the relative demand for unskilled workers and therefore account for observed deteriorating economic fortunes of unskilled workers in the UK.

This chapter empirically examines the roles of outsourcing typology in explaining its economic effects in Thailand's manufacturing sector. It aims to establish itself as an extension of previous analyses and discussions developed thus far. As in Görg and Hanley (2005), my establishment-level dataset enables me to separate materials outsourcing from services outsourcing. My empirical estimates reveal that outsourcing typology matters to changes in firm productivity, labour productivity and wage inequality.

More specifically, the estimates indicate that merely materials outsourcing triggers improvements of firm productivity whereas services outsourcing does not. The absence of firm productivity gains associated with services outsourcing may be attributed to managerial and operational inefficiency arising from contracting out of services, such as transactions costs, costs of contract enforcement and monitoring costs. Additionally, my evidence points to materials outsourcing as a sole contributor to a surge in labour productivity in Thailand's manufacturing sector, implying that compositional and structural changes concerned with contracting out of services hinder a firm to reallocate existing labour employed in-house to tap benefits from productivity gains from specialization. Furthermore, a breakdown of outsourcing activities into materials and services outsourcing provides a clearer insight into a linkage between production fragmentation and widening wage inequality. I find that materials outsourcing is a crucial driver of wage differentials between skilled and unskilled workers whereas the opposite effect is observed for services outsourcing. The contrasting effects on the relative demand for skilled workers are perhaps a root cause of weak empirical evidence I presented in Chapter 5.

The remainders of this chapter can be briefly structured as follows. Section 6.2 discusses outsourcing typology and its measurements. Section 6.3 portrays and analyses empirical evidence. I partition this section into three parts: outsourcing typology and firm productivity, outsourcing typology and labour productivity, and outsourcing typology and wage inequality. Section 6.4 concludes.

6.2 Outsourcing Typology

A brief survey of the literature that took into consideration outsourcing typology unanimously underscored biases and inaccurate results that arise from the uses of aggregate outsourcing indices. Disaggregation of the outsourcing notion is therefore essential. In this chapter, I adopt the first approach to disaggregating outsourcing activities — the nature of outsourced inputs — as in Görg and Hanley (2005), Görg et al. (2008) and Amiti and Wei (2009).

In particular, the indices of outsourcing intensity depart from the index of general outsourcing (O_i^T) employed thus far in the earlier chapters. I break down outsourcing activities into two types: materials outsourcing intensity (O_i^M) and services outsourcing intensity (O_i^S).

$$O_i^M = \frac{\sum_j MOUT_{ij}}{TC_i} \text{ and } O_i^S = \frac{\sum_k SOUT_{ik}}{TC_i} \qquad (6.1)$$

where $MOUT_{ij}$ denotes the uses of materials input j, including materials and component purchases, by an establishment i, and $SOUT_{ik}$ refers to the costs of services purchases k, including contract, commission work, repair and maintenance work, by an establishment i, and TC_i represents total production costs of an establishment i.

In principle, it is possible that various types of outsourced intermediate inputs may have different impacts on the relative demand for skilled workers. For instance, if services such as R&D and product designs are more skill-intensive than production of materials and components, outsourcing of services will reduce skill-intensive production activities and thus have a negative effect on the relative demand for skilled labour employed in-house. The opposite linkage is expected for a decision to contract out materials and components production. Therefore, to me it is worthwhile to further investigate the role of outsourcing typology such as materials and services outsourcing in the context of Thailand's manufacturing sector.

6.3 Empirical Evidence from Thailand's Establishment-Level Data

Outsourcing Typology and Firm Productivity

In Chapter 3, I presented empirical evidence that a firm pertaining to arm's length arrangements intensively, tends to have higher productivity in terms of output scaled by values of capital assets. The notion of outsourcing was broadly defined as purchases of intermediate inputs as in Girma and Görg (2004). In this chapter, I attempt to further investigate possibilities that different types of outsourcing have different effects on firm productivity.

Empirical evidence based on the establishment-level data indicates that outsourcing typology matters to its effects on firm productivity in Thailand's manufacturing industries. A fitted plot shown in Figure 6.1 demonstrates a positive, statistically significant relationship between materials outsourcing and firm productivity, implying that

FIGURE 6.1
A Fitted Plot of Materials Outsourcing Intensity and Output-to-Capital Ratio

$$y_{ki} = .5457^{***} + 2.8832^{***}O_i^T + u_i$$
$$(.1508) \qquad (.1849)$$
$$N = 6602 \; ; \; F = 234.18^{***} \; ; \; R^2 = .0355$$

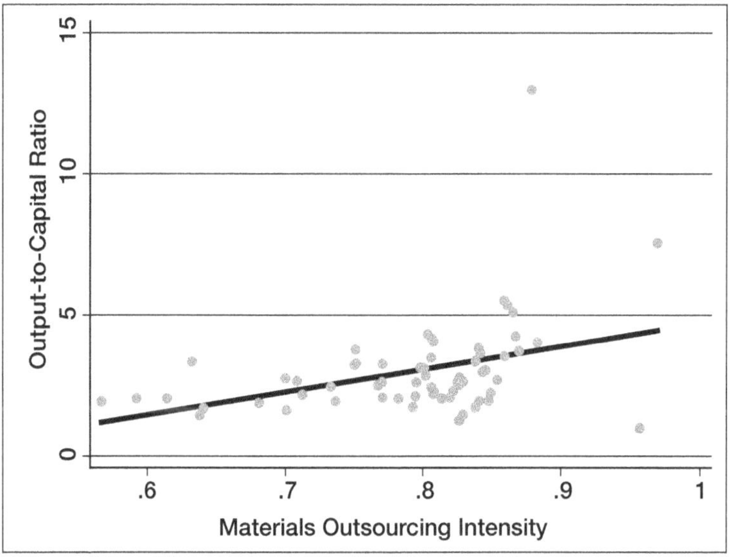

Notes: (1) *** represents statistical significance at 1 per cent; and (2) standard errors reported in parentheses.

contracting out materials production at arm's length enables a firm to reap benefits in terms of higher productivity from scale economies, distributing fixed costs and better access to greater varieties and higher quality of intermediate materials. In the long run, contracting out of materials production may also lead to changes in factor shares, which may have implications on firm productivity.

Outsourcing of services, in contrast, seems to exacerbate firm productivity. As revealed in Figure 6.2, the negative effects of services outsourcing on firm productivity may be attributable to added inefficiency and transactions costs associated with outsourcing activities, such as search and matching costs, monitoring costs, and transportation costs and so forth. My empirical findings that only materials

FIGURE 6.2

A Fitted Plot of Services Outsourcing Intensity and Output-to-Capital Ratio

$$y_{ki} = \underset{(.0680)}{3.4649^{***}} - \underset{(.5451)}{4.8167^{***}} O_i^S + u_i$$

$N = 4787$; $F = 78.08^{***}$; $R^2 = .0161$

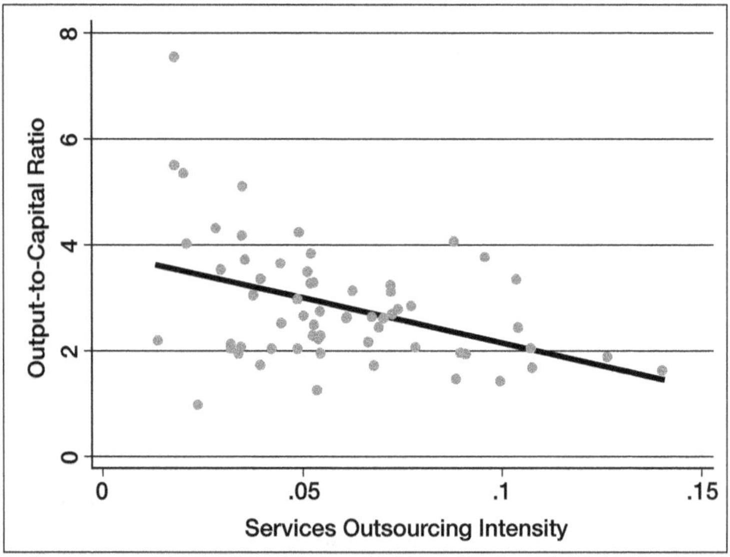

Notes: (1) *** represents statistical significance at 1 per cent; and (2) standard errors reported in parentheses.

outsourcing help improve firm productivity are strikingly consistent with those of Görg and Hanley (2005) which are confined specifically to international outsourcing.

A main message my empirical exercise in this section highlights is that the effects of outsourcing on firm productivity in Thailand's manufacturing industries depend crucially upon types of production activities involved. In addition, now that I have demonstrated that outsourcing typology matters to productivity effects of outsourcing, aggregation of the outsourcing notion as conventionally adopted by much previous research tends to produce biased estimates.

Outsourcing Typology and Labour Productivity

Having investigated how the nature of outsourcing activities accounts for the effects on firm productivity, I then focus on the roles of outsourcing typology in explaining changes in labour productivity. It should also be highlighted that the empirical exercise developed in this section establishes itself as an extension of that enumerated in Chapter 4.

Figure 6.3 portrays a fitted plot of materials outsourcing intensity and labour productivity. The estimates reveal that labour productivity measured by the ratio of output to total labour is positively and significantly correlated with materials outsourcing intensity.

However, there is no such effect observed for services outsourcing. Figure 6.4 depicts a relationship between services outsourcing intensity and labour productivity. The estimates show that the effects of services outsourcing on labour productivity are negative and statistically

FIGURE 6.3

A Fitted Plot of Materials Outsourcing Intensity and Labour Productivity

$$y_{li} = 8.9892^{***} + 3.4578^{***} O_i^M + u_i$$
$$(.1005) \qquad (.1220)$$
$$N = 8811 \; ; \; F = 803.57^{***} \; ; \; R^2 = .0836$$

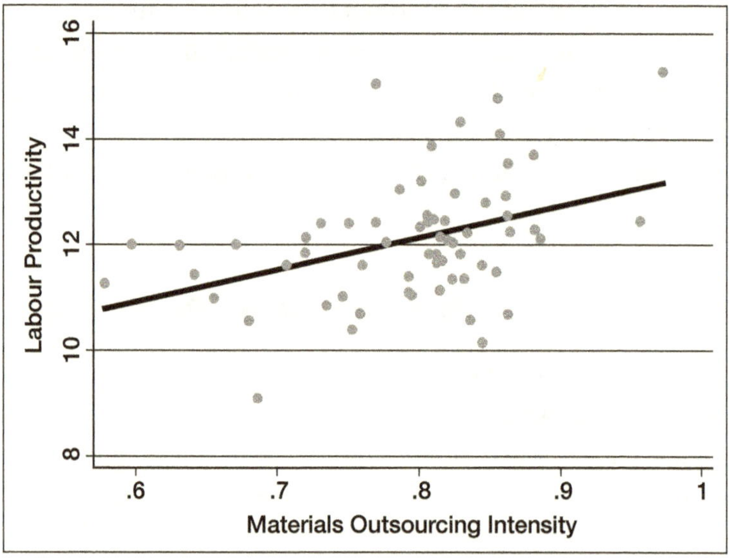

Notes: (1) *** represents statistical significance at 1 per cent; and (2) standard errors reported in parentheses.

significant at 1 per cent. The contrasting impacts on labour productivity may be explained by the fact that structural and compositional changes associated with services outsourcing hinder a firm to downsize its relatively inefficient stages of production and hence block away labour productivity improvements from more specialization. In a nutshell, outsourcing typology also plays a pivotal role in harnessing

FIGURE 6.4
A Fitted Plot of Services Outsourcing Intensity and Labour Productivity

$$y_{li} = 12.8242^{***} - 6.9147^{***}O_i^S + u_i$$
$$(.0391) \qquad (.2929)$$
$$N = 5952 \; ; \; F = 557.37^{***} \; ; \; R^2 = .0857$$

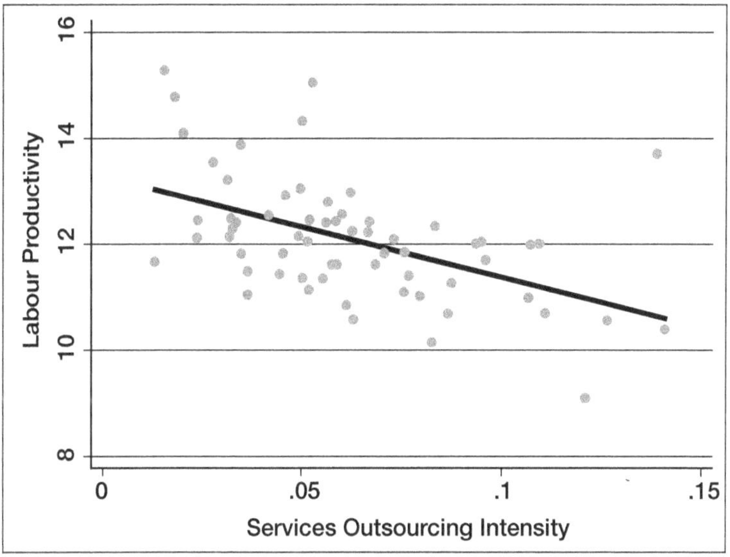

Notes: (1) *** represents statistical significance at 1 per cent; and (2) standard errors reported in parentheses.

changes in labour productivity in Thailand's manufacturing industries, and the empirical analyses that fail to account for outsourcing typology are likely to yield biased results.

Recall that in Chapter 5 the effects of the aggregate outsourcing intensity on wage inequality are rather ambiguous. A brief survey of literature presented earlier in this chapter points to outsourcing typology that plays a major role in the economic impacts of outsourcing activities. Accordingly, this section attempts to push forward the empirical investigations by segregating outsourcing activities into different types of fragmented activities that may exhibit different effects on wage inequality in Thailand's manufacturing sector.

Figure 6.5 portrays a fitted plot of the materials outsourcing intensity (o_i^M) and the wage shares of skilled workers (WS_{Hi}). It exhibits a slightly positive relationship between materials outsourcing activities and the relative demand for skilled workers, and the estimates further reveal that such a positive linkage is statistically significant though only at the 10 per cent. This implies that manufacturers in Thailand's manufacturing sector tend to contract out low skill-intensive materials production and thus carry out production of high skill-intensive materials in-house, thereby prompting an outward shift in the relative demand for skilled workers.

Outsourcing of services affects the wage shares of skilled workers in an opposed way. As shown in Figure 6.6, the correlation between the services outsourcing intensity and the wage gap between skilled and unskilled workers is negative and statistically significant at 5 per cent. Such a negative relationship may be rationalized by the extent to which the outsourced services like repair and maintenance work require skilled workers with specialized expertise and training, and therefore an establishment which contracts out services intensively at arm's length is likely to have lower wage shares for skilled workers.

The contrasting effects of materials and services outsourcing suggest that outsourcing typology matters to

FIGURE 6.5
A Fitted Plot of the Materials Outsourcing Intensity and the Shares of the Skilled Worker Payroll

$$WS_{Hi} = \underset{(.0032)}{.2814^{***}} + \underset{(.0017)}{.0032^{*}} O_i^M + u_i$$

$$N = 3998 \; ; \; F = 3.66^{*} \; ; \; R^2 = .0009$$

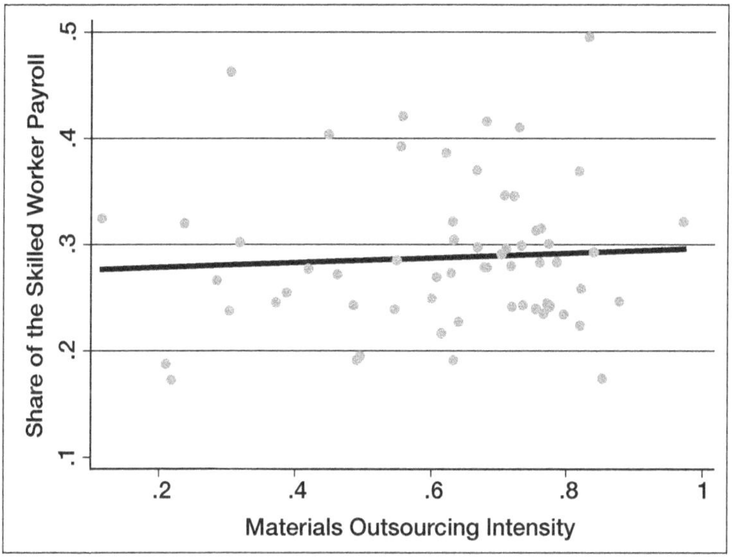

Notes: (1) *** represents statistical significance at 1 per cent; (2) * represents statistical significance at 10 per cent; and (3) standard errors reported in parentheses.

the analyses of wage inequality in Thailand's manufacturing sector. Again, aggregation of outsourcing activities as did in the last chapter as well as most existing studies would ultimately lead to biased results.

However, my empirical findings based on Thailand's establishment-level data seem to differ noticeably from those of Chongvilaivan et al. (2009) who employed the

FIGURE 6.6
A Fitted Plot of the Services Outsourcing Intensity and the Shares of the Skilled Worker Payroll

$$WS_{Hi} = .2615^{***} - .0050^{**}O_i^S + u_i$$
$$(.0085) \qquad (.0020)$$
$$N = 3459 \; ; \; F = 5.94^{**} \; ; \; R^2 = .0017$$

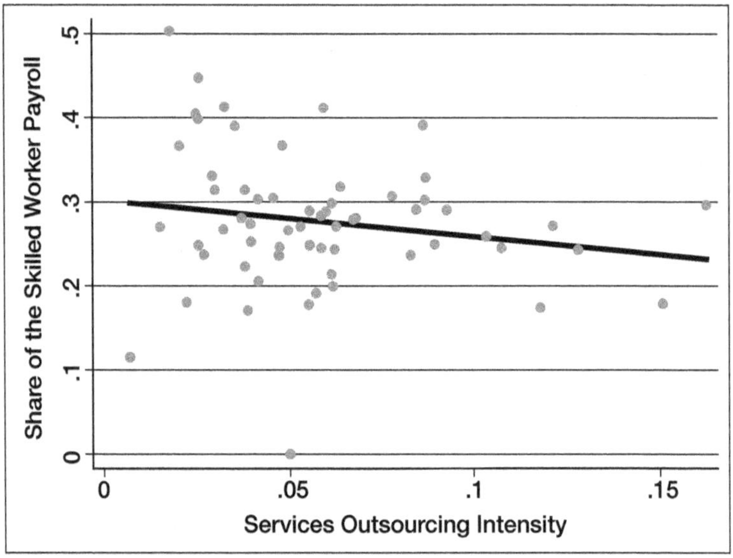

Notes: (1) *** represents statistical significance at 1 per cent; (2) ** represents statistical significance at 5 per cent; and (3) standard errors reported in parentheses.

U.S. manufacturing industry data disaggregated at the six-digit NAICS level and showed the opposite results that services outsourcing (and downstream materials outsourcing like reprocessing, repackaging and blending) is a key contributor to rising wage differentials between skilled and unskilled workers whereas upstream materials outsourcing like intermediate parts and components is not.

In this regard, Thai manufacturers' outsourcing behaviours depart from American manufacturers'. Based on my empirical findings, Thai manufacturers tend to contract out low skill-intensive parts and components and high skill-intensive services while American manufacturers incline to do the opposite. The negative effects of services outsourcing observed in Thailand's manufacturing sector may be explained by the argument that outsourcing of the skill-intensive services like information technology and human resource management — which are likely to be highly skill-intensive — has become part and parcel of Thai manufacturers (Paisittanand and Olson 2006).

6.4 Concluding Remarks

This chapter is essentially an extension of the empirical investigations developed thus far with a specific focus on the nature of outsourcing activities. Following Görg and Hanley (2005) and Görg et al. (2008), I partition the index of outsourcing intensity into two categories according to its tangibility: materials outsourcing intensity (o_i^M) and services outsourcing intensity (o_i^S). Empirical evidence is strikingly robust and reveals that the economic effects of outsourcing in Thailand's manufacturing industries rely crucially on its typology.

Firstly, my empirical exercise presented in this chapter sheds further light on the impacts of outsourcing on firm productivity. The estimates show that contracting out of intermediate inputs may not always enhance firm productivity. A firm tends to reap benefits from productivity improvements when undertaking materials outsourcing. However, the productivity effects are reversed for contracting out of services. The deterioration of firm productivity associated with a decision to contract out services activities at arm's length may be caused by arising inefficiency such as costs of searching appropriate suppliers, transaction costs, and monitoring costs.

Secondly, outsourcing typology has a major role to play in explaining a shift in labour productivity. My empirical estimates based on Thailand's establishment-level data manifest that materials outsourcing contributes to labour productivity improvements whereas services outsourcing does not. To me, the negative labour productivity effects of services outsourcing are surprising since trimming in-house production stages in general should enable the existing labour to specialize particular production activities and ultimately to enjoy higher productivity. The adverse effects of services outsourcing on labour productivity may be rationalized by compositional and structural changes within a firm that hold back the labour productivity enhancement.

Last but not least, a breakdown of outsourcing activities has made evidence of a relationship between outsourcing and skill upgrading discussed in Chapter 5 more clear-cut. I find that materials outsourcing accounts for an outward shift in the relative demand for skilled workers and thus contributes to widening wage inequality in Thailand's manufacturing sector while the opposite correlation is observed for services outsourcing. The contrasting impacts of materials and services outsourcing can be explained by skill intensity differentials. Since contracting out of materials like parts and components by definition tends to be less skill-intensive than that of services like repair and maintenance, a decision to downsize materials (services) production trims less (more) skill-intensive production stages carried out in-house, thereby triggering a shift (decline) in the relative demand for skilled workers.

Note

1. Chongvilaivan et al. (2009) also contemplated outsourcing of services in addition to outsourcing of upstream and downstream materials. They found that services outsourcing has a positive effect on the relative demand for skilled workers as does downstream materials outsourcing.

7
Policy Implications

7.1 Industrial Productivity Development

Since the last two decades, the formation of production fragmentation and networks in Southeast Asia has provided a wide array of business opportunities for local firms and Multinational Enterprises (MNEs) to leverage on ever-increasing productivity growth, especially in Malaysia, Singapore, Thailand and Vietnam. Based on this firm operating model, production stages are finely partitioned into many stages and carried out in different locations across regions and economies. As discussed in Chapter 3, there are at least two key mechanisms through which the prevalent use of production fragmentation accounts for a surge in firm productivity.

The first driver lies with vertical division and specialization of production factors, known as productivity gains from specialisation, and is attributable to technological advancement like computer and electronic revolutions in business management and operations. Thailand's machinery industry, for instance, participated in production and distribution networks by producing parts and components and supplying them to the growing mobile cluster industry. In this sense, outsourcing of parts and components allows the automobile and machinery industries to carry out what they are good at and ultimately enhance their productivity and competitiveness as a whole.

The other catalyst of productivity development is concerned with the improved access to intermediate inputs which are available at lower costs and/or better quality than what outsourcing firms had obtained by in-house production. This driving force of productivity growth emanates mainly from technology transfers and spillovers that help defray costs, expand varieties, and improve quality of intermediate inputs available to them. Furthermore, in comparison with in-house production, contracting out of production stages at arm's length offers much larger adjustability and resilience of production patterns and thus much more opportunities for firms to reap benefits from technology transfers or spillovers, especially the physical movements of technology and managerial know-how.

The latter productivity-enhancing mechanism is particularly vital to developing Southeast Asia as recent developments of production fragmentation and networks have presented new challenges of technology transfers and spillovers across industries and regions. Under the competitive business environment, arm's length transactions that entail industrial agglomerations and vertical division of production factors boost firms' absorptive capacity of technology — a crucial element of technological spillover process. It would put pressure on local firms and MNEs to actively seek productive business partners and/or suppliers to secure their sources of intermediate parts and components at acceptable quality, specifications, prices and delivery timing. However, participation in production networks, as do other arm's length transactions, pertains to the inefficiency arising from costs of transactions and structural adjustments, at least in the short run. In the traditional sector like garment and textile, downsizing production stages typically brings about huge adjustment costs due to their relatively low speed and low frequency of transactions. The costs of structural adjustments tend to be lower in more advanced sectors like machinery and electronics industries.

My empirical evidence presented in Chapters 3 and 7 showed that outsourcing is a double-edged sword to industry productivity growth. In Chapter 3, I found that outsourcing of intermediate inputs in general prompts a surge in firm productivity. An extension presented in Chapter 7, however, further reveals that its typology has a large role to play in firm productivity gains from joining production networks. The implications drawn from my empirical results are that outsourcing activities may not always be beneficial to firm productivity. Therefore, well-tailored industrial policy to harness the effects on firm productivity and alleviate the inefficiency costs incurred in the process of production fragmentation is highly indispensable and must be put in place; the smooth process of industrial productivity development is put at risk otherwise.

Production fragmentation has put forward policy options to expedite the process of industrialization, alternative to traditional development strategies — in particular, import-substituting and export-oriented industrialization — which are unlikely to provide much boost in industrial productivity to most Southeast Asian economies. The reasons why the productivity effects of these industrialization strategies are largely limited are as follows. First of all, the trade restrictions imposed for import-substituting sector promotion tend to exacerbate the competitive business environment, thereby discouraging, if not deterring, MNEs to carry out technology transfers to local firms. Neither does export-oriented industrialization prompt an incentive for MNEs to transfer technology and managerial know-how to the region in that they typically relocate less skill-intensive, lower value-added production to developing countries where unskilled labour is abundant, while more skill-intensive, higher value-added production activities remain undertaken in developed countries.

As production fragmentation has become an increasingly important source of productivity growth in Southeast Asia, outsourcing and subcontracting offer new business opportunities local firms can leverage on. These pertain to a prevalent use of information and communication technology, particularly computers and internet, which allows a firm to be fully integrated into the vertical chain of production. For instance, the information technology revolution has tremendously reduced transaction costs, offered flexibility in doing business, and ultimately become a new source of comparative advantages in Singapore and Hong Kong, in which firms are able to tap benefits from the information technology development, expand their business operations and opportunities, and obtain more flexible arm's length arrangements.

Production fragmentation, in contrast, sets a stage for technological transfers and spillovers. Outsourcing enables local firms to climb up the value chain of production and improve their sophistication of in-house production activities. In addition, it also buttresses competitive business environment which makes MNEs more willing to transfer their technology and know-how. In this regard, production fragmentation offers Southeast Asia's policy-makers an alternative policy option to enhance industrial productivity and further the process of industrialization.

7.2 Labour Market Development

The potential negative impacts of production fragmentation on labour productivity, wages and employment, together with an enormous amount of attention in the media and political circles in recent times, are perhaps the main reason why much research has been devoted to relating the phenomenon and economic implications of outsourcing to labour markets. There is a growing realization that production fragmentation may be just as important, if not even more important, than technological progress to labour market effects.

Even though most of the current literature concerned with the economic impacts of production fragmentation on labour markets put much emphasis on developed economies, technical advances in information technology substantially lowering transportation and transaction costs have prompted surges in outsourcing activities in various developing countries like China and the key Association of Southeast Asian Nations (ASEAN) countries of Indonesia, Malaysia, the Philippines and Thailand. Contracting out production stages at arm's length has also been commonplace in various industries. For instance, in the plastic industry the R&D activities are typically contracted out due to the lack of intra-firm technology and human capital.

The textile and fashion industries are also outsourcing marketing and packaging activities to outside contractors to tap cost-saving benefits. The prevalence of outsourcing activities, nevertheless, gives rise to growing concerns of structural adjustments in domestic labour markets. An example can be found as the conflict between Thai Airways International Public Company Limited and its labour union (*Bangkok Post*, 11 February 2005). Thai Airways' decision to downsize its human resource department triggered a protest as the use of external recruitment agencies spurred the fear of job losses among 5,200 unionized employees.

Empirical evidence presented in Chapters 4–6 based on Thailand's establishment-level data reveals that production fragmentation may not always be detrimental to labour market development. In particular, an empirical investigation of the linkages between outsourcing activities and labour productivity demonstrates robust evidence that a decision to contract out production activities essentially accounts for a surge in labour productivity and employment in Thailand's manufacturing industries. Furthermore, even though in Chapter 5 I found merely weak evidence of the impacts on wage differential between skilled and unskilled workers, further investigations developed in Chapter 6 indicated that only outsourcing of intermediate materials contributes to the widening wage gap, whereas that of intermediate services does not. In this sense, empirical exercises carried out in Chapter 6 had pushed forward the roles of outsourcing typology as a key determinant of skill-biased shifts in wages.

As the manufacturing sector in the Southeast Asian countries are moving toward more capital- and technology-intensive activities, the impacts of materials and services outsourcing will have indispensable implications on skill-upgrading policy, also on the rising wage inequality between skilled and unskilled workers in the economies. The potential effects of production fragmentation are associated with labour productivity improvements. Nevertheless, it may also deteriorate wage differentials across skill groups. Therefore, the governments have a crucial role to play in alleviating the adverse effects without sacrificing potential labour productivity gains from outsourcing production. Training and skill upgrading of labour will be essential to move unskilled workers to more productive, more value-added sectors in the economy. To this end, improvement and upgrading of education as well as innovation programmes in the regional economies serve as important driving forces of augmented economic benefits, which must be underway.

My empirical analyses pointed to winners and losers from production fragmentation as the estimates suggested that materials outsourcing tends to be in favour of skilled workers whereas services outsourcing seems to be biased towards the unskilled group. This may have vital policy implications on the compensation schemes that aim to bail out the losers and to ease adjustment costs associated with rising production fragmentation, including fair labour market regulations, unemployment benefits and apposite unionization, among others.

Bibliography

Abraham, K.G. and S.K. Taylor. "Firms' Use of Outside Contractors: Theory and Evidence". *Journal of Labor Economics*, vol. 14, no. 3 (1996): 394–424.

Ahn S., K. Fukao, and K. Ito. "Outsourcing in East Asia and Its Impact on the Japanese and Korean Labor Markets". OECD Trade Policy Working Papers no. 65. OECD, 2008.

Amiti, M. and J. Konings. "Trade Liberalization, Intermediate Inputs, and Productivity: Evidence from Indonesia". *American Economic Review*, vol. 97, no. 5 (2007): 1611–38.

Amiti, M. and S-J Wei. "Service Offshoring and Productivity: Evidence from the US". *The World Economy*, vol. 32, no. 2 (2009): 203–20.

Anderton, B. and P. Brenton. "Outsourcing and Low-skilled Workers in the UK". *Bulletin of Economic Research*, vol. 51, no. 4 (1999): 267–85.

Arndt, S.W. "Globalization and the Open Economy". *North American Journal of Economics and Finance*, vol. 8, no. 1 (1997): 71–79.

Arndt, S.W. and H. Kierzkowski. *Fragmentation: New Production Patterns in the World Economy*. Oxford: Oxford University Press, 2001.

Calabrese, G. and F. Erbetta. "Outsourcing and Firm Performance: Evidence from Italian Automotive Suppliers". *International Journal of Automotive Technology and Management*, vol. 5, no. 4 (2005): 461–79.

Chongvilaivan, A. and J. Hur. "Outsourcing, Labor Productivity and Wage Inequality in the US: A Primal Approach". *Applied Economics*, vol. 43, no. 4 (2011): 487–502.

Chongvilaivan, A., J. Hur, and Y.E. Riyanto. "Outsourcing Types, Relative Wages, and the Demand for Skilled Workers: New Evidence from U.S. Manufacturing". *Economic Inquiry*, vol. 47, no. 1 (2009): 18–33.

Deardorff, A.V. "Fragmentation in Simple Trade Models". *North American Journal of Economics and Finance*, vol. 12, no. 2 (2001): 121–37.

Dell'mour, R., P. Egger, K. Gugler, and M. Pfaffermayr. "Outsourcing of Austrian Manufacturing to Eastern European Countries: Effects on Productivity and the Labor Market". In *Fragmentation of the Value-Added Chain*, edited by S. Arndt, H. Handler, and D. Salvatore. Vienna: Austrian Ministry for Economic Affairs and Labour, 2000.

Díaz-Mora, C. "What Factors Determine the Outsourcing Intensity: A Dynamic Panel Data Approach for Manufacturing Industries". *Applied Economics*, vol. 40, no. 19 (2008): 2509–21.

Egger, H. and J. Falkinger. "The Distributional Effects of International Outsourcing in 2x2-Production Model". *North American Journal of Economics and Finance*, vol. 14, no. 2 (2003): 189–206.

Egger, H. and P. Egger. "International Outsourcing and the Productivity of Low-skilled Labour in the EU". *Economic Inquiry*, vol. 44, no. 1 (2006): 98–108.

Fajnzylber, P. and A.M. Fernandes. "International Economic Activities and Skilled Labour Demand: Evidence from Brazil and China". *Applied Economics*, vol. 41, no. 5 (2009): 563–77.

Feenstra, R.C. and G. Hanson. "Foreign Investment, Outsourcing and Relative Wages". In *Political Economy of Trade Policy: Essays in Honor of Jagdish Bhagwati*, edited by R.C. Feenstra, G.M. Grossman, and D.A. Irwin. Cambridge: MIT Press, 1996*a*.

———. "Globalization, Outsourcing, and Wage Inequality". *American Economic Review*, vol. 86, no. 2 (1996*b*): 240–45.

———. "The Impact of Outsourcing and High-technology Capital on Wages: Estimates for the United States". *Quarterly Journal of Economics*, vol. 114, no. 3 (1999): 907–40.

Fuss, M., D. McFadden, and Y. Mundlak. "A Survey of Functional Forms in Economics Analysis". In *Production Economics: A*

Dual Approach to Theory and Applications, edited by M. Fuss and D. McFadden. Amsterdam: North Holland, 1978.

Geishecker, I. "Outsourcing and the Relative Demand for Low-Skilled Labour in German Manufacturing: New Evidence". German Institute for Economic Research DIW-Berlin Discussion Papers, no. 313, 2002.

Geishecker, I. and H. Görg. "Winners and Losers: A Micro-level Analysis of International Outsourcing and Wages". *Canadian Journal of Economics*, vol. 41, no. 1 (2008): 243–70.

Girma, S. and H. Görg. "Outsourcing, Foreign Ownership, and Productivity: Evidence from UK Establishment-Level Data". *Review of International Economics*, vol. 12, no. 5 (2004): 817–32.

Görg, H. and A. Hanley. "International Outsourcing and Productivity: Evidence from the Irish Electronics Industry". *North American Journal of Economics and Finance*, vol. 16, no. 2 (2005): 255–69.

Görg, H., A. Hanley, and E. Strobl. "Productivity Effects of International Outsourcing: Evidence from Plant-level Data". *Canadian Journal of Economics*, vol. 41, no. 2 (2008): 670–88.

Görzig B. and A. Stephan. "Outsourcing and Firm-level Performance". DIW Berlin Discussion Papers, no. 309, 2002.

Grossman, G.M. and E. Helpman. "Integration versus Outsourcing in Industry Equilibrium". *Quarterly Journal of Economics*, vol. 117, no. 1 (2002): 85–120.

Gujarati, D. and L. Dars. "Production and Nonproduction Workers in the U.S. Manufacturing Industries". *Industrial and Labor Relations Review*, vol. 26, no. 1 (1972): 660–69.

Hanson, G.H. and A.E. Harrison. "Trade, Technology, and Wage Inequality in Mexico". *Industrial and Labor Relations Review*, vol. 52, no. 2 (1999): 271–88.

Head, K. and J. Ries. "Offshore Production and Skill Upgrading by Japanese Manufacturing Firms". *Journal of International Economics*, vol. 58, no. 1 (2002): 81–105.

Hijzen, A. "International Outsourcing, Technological Change, and Wage Inequality". *Review of International Economics*, vol. 15, no. 1 (2007): 188–205.

Hijzen, A., H. Görg, and R.C. Hine. "International Outsourcing and the Skilled Structure of Labour Demand in the United Kingdom". *Economic Journal*, vol. 115, no. 506 (2005): 860–78.

Holmes, T.J. "Localization of Industry and Vertical Disintegration". *Review of Economics and Statistics*, vol. 81, no. 2 (1999): 314–25.

Hsieh, C-T and K.T. Woo. "The Impact of Outsourcing to China on Hong Kong's Labor Market". *American Economic Review*, vol. 95, no. 5 (2005): 1673–87.

Hummels, D., J. Ishii, and K-M. Yi. "The Nature and Growth of Vertical Specialization in World Trade". *Journal of International Economics*, vol. 54, no. 1 (2001): 75–96.

Jones, R.W. and H. Kierzkowski. "A Framework for Fragmentation". In *Fragmentation: New Production Patterns in the World Economy*, edited by S.W. Arndt and H. Kierzkowski. New York: Oxford University Press, 2001.

Levinson, J. and A. Petrin. "Estimating Production Functions Using Inputs to Control for Unobservables". *Review of Economic Studies*, vol. 70, no. 2 (2003): 317–42.

Machin S., A. Ryan, and J. van Reenen. "Technology and Changes in Skill Structure: Evidence from A Panel of International Industries". Centre for Economic Policy Research Discussion Paper, no. 1434 (1996).

Markusen, J.R. "The Boundaries of Multinational Enterprises and the Theory of International Trade". *Journal of Economic Perspectives*, vol. 9, no. 2 (1995): 169–89.

Olsen, K.B. "Productivity Impacts of Offshoring and Outsourcing: A Review". OECD, STI Working Paper, no. 1 (2006).

Paisittanand, S. and D.L. Olson. "A Simulation Study of IT Outsourcing in the Credit Card Business". *European Journal of Operation Research*, vol. 175, no. 2 (2006): 1248–61.

Schmitz, H. and P. Knorringa. "Learning from Global Buyers". *Journal of Development Studies*, vol. 37, no. 2 (2000): 177–205.

Siegel, D. "The Impacts of Computers on Manufacturing Productivity Growth: A Multiple-Indicators, Multiple-Causes Approach". *Review of Economics and Statistics*, vol. 79, no. 1 (1997): 68–78.

Siegel, D. and Z. Griliches. "Purchases Services, Outsourcing, Computers, and Productivity in Manufacturing". In *Output Measurement in the Service Sector*, edited by Z. Griliches. Chicago: University of Chicago Press, 1992.

ten Raa, T. and E.N. Wolff. "Outsourcing of Services and the Productivity Recovery in U.S. Manufacturing in the 1980s and 1990s". *Journal of Productivity Analysis*, vol. 16, no. 2 (2001): 149–65.

Thangavelu, S.M. and A. Chongvilaivan. "The Impact of Material and Service Outsourcing on Employment in Thailand's Manufacturing Industries". *Applied Economics*, 2010.

Wattanapruttipaisan, T. "SME Subcontracting as a Bridgehead to Competitiveness: Framework for an Assessment of Supply-Side Capabilities and Demand-Side Requirements". *Asia-Pacific Development Journal*, vol. 9, no. 1 (2002): 65–87.

Index

About the Author

Aekapol Chongvilaivan is a Fellow and Coordinator of Regional Economic Studies Programme at the Institute of Southeast Asian Studies (ISEAS), Singapore. Dr Aekapol's current research covers a wide range of economics areas such as fragmentation and outsourcing in Southeast Asia, international trade, regional economic integration, and applied econometrics, among others. His research has been published in several international refereed journals including *Economic Inquiry*, *Southern Economic Journal*, *Applied Economics*, *Journal of Institutional and Theoretical Economics*, and *The World Economy*.

www.ingramcontent.com/pod-product-compliance
Lightning Source LLC
Chambersburg PA
CBHW021103210326
41598CB00016B/1312
9789814311267